CAMBRIDGE LIBRARY COLLECTION

Books of enduring scholarly value

Life Sciences

Until the nineteenth century, the various subjects now known as the life sciences were regarded either as arcane studies which had little impact on ordinary daily life, or as a genteel hobby for the leisured classes. The increasing academic rigour and systematisation brought to the study of botany, zoology and other disciplines, and their adoption in university curricula, are reflected in the books reissued in this series.

The Fern Garden

James Shirley Hibberd (1825–90) was a journalist and writer on gardening, whose popular works had great influence on middle-class taste. Although not a trained horticulturalist, his many books were based on practical experience. He developed a succession of gardens in north London concentrating on particular types of plants, and his books reflected this work, with the *Rose Book* (1864) and the *Fern Garden* (1869) being particularly successful. He also wrote on garden design, flower arrangement, garden furniture and architecture, and his *Rustic Adornments* of 1856, also published in this series, is an important work of social and fashion history. He edited the magazine *Floral World* until 1875 and later the *Gardener's Magazine*, and was even consulted by the government about potato blight. His engaging and very personal style made him a popular forerunner of modern celebrity gardeners, and set a fashion for highly decorative and ornamental gardens.

Cambridge University Press has long been a pioneer in the reissuing of out-of-print titles from its own backlist, producing digital reprints of books that are still sought after by scholars and students but could not be reprinted economically using traditional technology. The Cambridge Library Collection extends this activity to a wider range of books which are still of importance to researchers and professionals, either for the source material they contain, or as landmarks in the history of their academic discipline.

Drawing from the world-renowned collections in the Cambridge University Library, and guided by the advice of experts in each subject area, Cambridge University Press is using state-of-the-art scanning machines in its own Printing House to capture the content of each book selected for inclusion. The files are processed to give a consistently clear, crisp image, and the books finished to the high quality standard for which the Press is recognised around the world. The latest print-on-demand technology ensures that the books will remain available indefinitely, and that orders for single or multiple copies can quickly be supplied.

The Cambridge Library Collection will bring back to life books of enduring scholarly value (including out-of-copyright works originally issued by other publishers) across a wide range of disciplines in the humanities and social sciences and in science and technology.

The Fern Garden

How to Make, Keep, and Enjoy it.

Shirley Hibberd

CAMBRIDGE UNIVERSITY PRESS

Cambridge, New York, Melbourne, Madrid, Cape Town,
Singapore, São Paolo, Delhi, Tokyo, Mexico City

Published in the United States of America by Cambridge University Press, New York

www.cambridge.org
Information on this title: www.cambridge.org/9781108037181

This edition first published 1869
This digitally printed version 2011

ISBN 978-1-108-03718-1 Paperback

The original edition of this book contains a number of colour plates,
which have been reproduced in black and white. Colour versions of these
images can be found online at www.cambridge.org/9781108037181

CYSTOPTERIS DICKIEANA.

THE FERN GARDEN

HOW TO MAKE, KEEP, AND ENJOY IT;

OR,

FERN CULTURE MADE EASY.

BY

SHIRLEY HIBBERD,

AUTHOR OF "RUSTIC ADORNMENTS FOR HOMES OF TASTE,"
ETC. ETC.

ILLUSTRATED WITH EIGHT COLOURED PLATES AND
FORTY WOOD ENGRAVINGS.

LONDON
GROOMBRIDGE AND SONS,
5, PATERNOSTER ROW.

MDCCCLXIX.

LONDON; PRINTED BY J. E. ADLARD, BARTHOLOMEW CLOSE.

PREFACE.

Beginners in Fern culture are very much perplexed by the abundance of books on the subject, and their general unfitness to afford the aid a beginner requires. Almost everybody has written a book on ferns, it having become the fashion to consider a knowledge of the subject rather a disqualification than otherwise. When the blind attempt to lead the blind the result can be safely predicated, and no doubt the myriads of bad books on ferns that swarm in the cheap book shops have done their full share of mischief. We have fortunately plenty of good books on the subject, but for the most part they are technical and elaborate, and shoot over the heads of beginners. Some of my fern-loving friends have persuaded me to try my hand on a small volume adapted for the induction of the unlearned and unskilled in this pursuit, and here it is. Whether it will supersede any of the bad books or take lowest rank amongst them is for me a solemn problem. But I send it forth in hope that after a quarter of a century of hard work in the practical part of the subject, I may be better qualified to make a little book than some of those who, previous to writing, had acquired only a week or so of experience, and a very dim knowledge of about half a dozen species. As almost every fern in cultivation has names enough to fill a small volume, I have in every case adopted the names by which those recommended are best known in nurseries and gardens. The fearful question of nomenclature is thus avoided, and every fern may be found by the name it bears in this epitome.

S. H.

CONTENTS.

THE FERN GARDEN:

HOW TO MAKE, KEEP, AND ENJOY IT;

OR,

FERN CULTURE MADE EASY.

CHAPTER I.

FERNS IN GENERAL.

I HAVE a fine opportunity now for a dry chapter. I have a good mind to hang up a tuft of straw to indicate that the way is dangerous, and to warn the reader not to proceed a line further. Ferns, my friends, belong to the sub-kingdom of vegetables termed CRYPTOGAMIA, a sub-kingdom so named because it is the custom of the population to celebrate marriages in the dark, so that it can scarcely be averred of them to a certainty that they really marry at all. In this sub-kingdom there are several large tribes, such as the mosses, the horse-tails, the lichens and liver worts; but the ferns or *filices* are the most noble of all, associating with others freely, but towering above them in apparent consciousness of right to rule.

All the cryptogams are destitute of flowers; that is one of their most noticeable distinctions. But though flowerless they, for the most part, produce seeds in plenty. Look on the under side of a ripe frond of almost any fern you can get hold of, and you will observe sharp lines, or dots, or constellations of red, brown, or yellow fruit or spore cases; within these are the *spores* or true seeds, by the germination of which the race is multiplied.

Ferns differ from flowering plants in the principles of their construction and growth. If we examine the base of a leaf-stalk of a tree we shall find a bud there, which, if left alone, will produce a branch or a cluster of fruit the next season. There are no such buds in the axils of fern leaves, not even in those of the brake, which is peculiarly tree-like in its growth. The growth of a fern is a sort of perpetual lengthening out at both ends. The upward growth, which is more frequently the subject of observation than the growth of the roots, consists first in a process of unrolling, and then of expansion and maturation of the leaves and stems. Because of these and other characters which obviously and without reference to the peculiar nature of their fruit distinguish them from flowering plants, the several parts of a fern are named differently to the corresponding parts in flowering plants. Thus, the true stem or root-stock of a fern is called a *caudex,* the true leaf is called a *frond,* the stem which bears the leaf is called the *stipes,* and the ramifications of the stipes through the leafy portion corresponding to the leaf-stalks of other plants bears the name of *rachis.* These are all the technicalities we need be troubled with, save and except as we go on the names of the ferns themselves. From the sublime to the ridiculous is but a step. I have just made that step while walking through the fern-house to obtain the needful inspiration to write this little book. There I saw my plumy emerald green pets glistening with health and beadings of warm dew, and I thought it might help me if I read their names. Here are a few of them—Acrostichum Requienianum, Alsophila Junghuhniana, Anemia Schimperiana, Aspidium Karwinskyanum, Polystichum Plaschnichianum, Asplenium Gaudichandianum, Euphegopteris hexagonopterum, Dictyopteris megalocarpum. You must endure this sort of thing if you purpose giving the slightest amount of attention to ferns, for only a few out of thousands have English names, and to translate the botanical names into English would be very imprudent, not to say sometimes impossible. But I assure you the names do not spoil the plants, they only compel fern books to be ugly and forbidding. Carolina Wilhelmina Amelia Skeggs was an unamiable person, but my Mohria thurifraga var. Achilliæfolia is as sweet a bit of vegetable jewellery as you are likely to meet with in a day's march, and I am sure you will admire, when you find it, Didymoglossum vel Trichomanes radicans.

CHAPTER II.

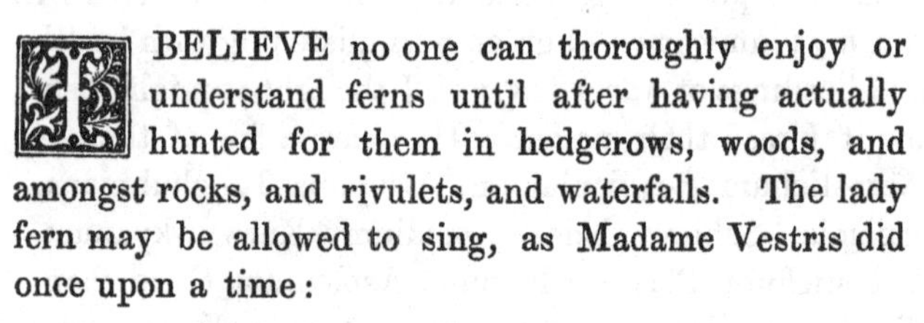

I BELIEVE no one can thoroughly enjoy or understand ferns until after having actually hunted for them in hedgerows, woods, and amongst rocks, and rivulets, and waterfalls. The lady fern may be allowed to sing, as Madame Vestris did once upon a time:

> Through the woods, through the woods,
> Follow and find me,
> Search every hollow, and dingle, and dell,
> I leave but the print of my footstep behind me;
> So those who would find me must search for me well.

I cannot afford space to enlarge upon the joys of fern-collecting, the pic-nicing, archæologico-exploring, and holiday perambulating that may be associated with the sport. Kindly imagine all this and save me the expenditure of space on anything but the business in hand. Ferns are so widely distributed that wherever a rural walk is possible, it is almost certain that somewhere in the district ferns may be found. The south-western counties of England constitute the home paradise of the fern collector, but, as we must find our happiness where our lot is cast, it is better to make the most of the ferns within our reach than to repine if Cornwall and Devon happen to be *terra incognita*. In the neighbourhood of London are many localities rich in ferns, but as these are for the most part pretty well known I shall not enumerate them, but proceed at once to make some remarks on collecting ferns for cultivation. It is only during the height of summer that the deciduous kinds can be readily found by inexperienced collectors, and it is at that season that fern hunting proves a particularly agreeable pastime. It would be better always if the ferns could be removed from their native sites when first about to commence their new growth in the spring, and this can be done sometimes by searching in woods and hedgerows for old fronds, and tracing them to their source. The roots should then be taken up without injury to the crowns, and be at once planted or potted as required, and assisted with shade and shelter until established in the places assigned them in the garden. Experienced collectors may hunt for ferns during the winter to great advantage in districts where they are known to abound, as in the event of a mild season many of the deciduous kinds will be still green; and evergreen kinds, such as hartstongue and common polypody, may be better lifted in winter than at any other season. But as a rule fern hunting is a recreation for summer time, and any fern may be taken up in the height of summer and be kept with the utmost certainty for cultivation; the worst that is likely to happen is the loss of all the fronds they carry at the time of taking up; but a new crop will soon succeed them if proper care be taken. The fern collector should be provided with aids and implements adapted to the county in which he is about to make explorations. Where only terrestrial and hedgerow kinds are expected to be found, a large basket, or better, a pair of baskets of moderate size, such as can be carried one in each hand, will be necessary. They should have close fitting lids, because if ferns are taken up on a hot day and exposed for some hours to the atmosphere, the crowns and roots will be so much exhausted that some may die, and all will be injured, whereas by packing them close with a little moist moss amongst them, the roots and crowns will be kept tolerably fresh until they can be potted or planted out. A short-handled three-pronged fork and a trowel, and a strong clasp knife, will be needful; and in some instances it will be necessary to borrow a spade or digging fork near the spot where operations are to take place, for fine old roots of royal osmund and other large-growing ferns will defy the leverage of all small hand tools. When ferns of large size are taken up in the height of summer, *it is best to cut away all or nearly all their fronds at once, and use those fronds as packing material.*

On reaching home, the best treatment to subject them to is to pot them all separately in the smallest pots their roots can be got into, with cocoa-nut fibre alone or the fibre of good peat or leaf-mould, and shut them up in a frame, and keep only moderately moist until they start into growth. As at this early stage of the study I may suppose you do not know how to pot them and restore their energies, I will endeavour to point out a simpler mode of procedure. Find a very shady place in the garden and there make a bed of leaf mould or peat soil, or cocoa-nut fibre refuse, and plant the ferns in it as close together as possible. Then cover them with bell glasses or common hand lights, and sprinkle them with water every evening, but take care not to make them very wet at the roots. They will soon begin to grow. In the spring following you may plant them in the fernery.

Small ferns found growing on rocks and walls must always be carefully dealt with. The little maidenhair spleenwort will sometimes send its black wiry roots quite through the substance of a nine-inch or fourteen-inch wall, and to remove it with complete roots is then quite out of the question. By loosening a portion of its hold just below the crown of the plant, roots may generally be obtained sufficient to enable it to re-establish itself under cultivation. A strong chisel and a hammer will be required in undertakings of this sort, and it may be well to add a little discretion also, especially as to extent to which walls—the property of somebody—are to be injured for the sake of a tuft of fern worth but a few pence, and of which specimens may be obtained more easily by further search without any necessity for the infliction of damage. Ferns found growing on and amongst rocks should always, if possible, be obtained with portions of the rock to which they are attached. If this cannot be accomplished, carefully tear the plant from the rock in a way to injure the roots as little as possible; good pieces will soon emit roots and fronds if properly treated, especially if kept moist by packing in moss or sphagnum from the first moment of obtaining the specimen. Allow me to remark, further, that the passion for fern collecting has in many instances been carried to a ridiculous excess by persons who merit the title not of fern collectors so much as fern destroyers. Let every genuine lover of ferns be on his guard both to discourage reckless fern collecting, and protect as far as possible the few remaining localities of scarce British ferns. It is not many years since I saw amongst a heap of dried mosses, ferns, grasses, &c., in the possession of a lady, a *sheet* of Tunbridge fern nearly a yard square. This had been torn from its native site, carefully rolled up like a piece of old blanket, and put away, and was afterwards brought forth as a trophy, and preserved as a memorial of the days "when we went gipsying." The value of that sheet when fresh might have been about £5, and no doubt any nurseryman could make a larger sum of a good square yard of the Tunbridge fern. Such reckless destruction, such base contempt for the value set upon a rare fern by those who understand its history and its habits, and appreciate the interest that arises out of its beauty and rarity combined, is to be considered as a crime; and though there is no law to punish the perpetrator, except in cases where there might be an action for trespass or wilful damage, it is the duty of every conservator of our native flora to visit crimes of this kind with the sternest disapprobation, accompanied with truthful explanations of the injury done alike to natural scenery and to science by such acts of spoliation.

If you can dig up ferns in early spring, you may plant them in your fernery at once, and if shaded for a time and frequently sprinkled with water, taking care always not to make the soil about them very wet, they will soon begin to grow vigorously, and after that patience is the only quality required on your part to ensure your proper reward.

You will soon learn to distinguish ferns from all other plants when you meet with them. When you find a fern, take notice of the soil and situation it is growing in, and in attempting its cultivation imitate those conditions as nearly as possible. The pretty wall rue spleenwort loves to grow in the full sun, upon and amongst sandstone rocks. You will see plenty of it on the approaches to the Suspension Bridge at Clifton, and you may find the common maiden-hair spleenwort keeping it company if you look sharp. It is in the shady, dank, almost dripping hollow, or on the slope of a water-course, that you are most likely to find the lovely lady fern, the hard fern, and the royal osmund, yet these will sometimes make a bonny show upon dry banks beside a dusty highway, where, perhaps, for miles the common lastrea is the prevailing fern of the district. In Epping Forest there are thousands of pollard trees on the awkward stems of which are perched, like wreaths of honour, tufts of the common polypody. I used when a boy to tear them off to line my basket with when birdnesting, for that forest was my playground. If you want to see the bracken you need not travel far, but if you would cultivate it you must notice that it grows to its grandest stature on mellow, yellowish loam, and is rather poor and stunted on sand and peat, though not always so. Observe always how they look when they are at home, and thereby learn to persuade them to believe themselves at home when you have planted them in the garden. Some thrive on perpendicular walls of stone and brick, others in the moist woodland shade, others on the bleak mountain top, and many a glorious group may be found on the sides and roofs of caverns, which they make like fairy palaces with their green feathery plumes and golden dottings of mysterious fruit. However many lessons you may learn of the habits of the several kinds of ferns, there should be one lesson impressed upon your mind more deeply than any—it is this, that, much as they love moisture, it is a most rare thing to see a fern growing with its roots naturally in water. When they congregate, as it were, to drink of the brook that passes by, they keep their feet clear away from the current, and lodge safely on the slopes that dip towards the water; or stand proudly upon little islets that compel the stream to sing as it passes them; or on banks and hummocks round about where they can enjoy the tiny splashes the trout make when they leap for flies, and the soft nourishing vapour that rises day and night amongst their shining fronds. Yes, it is upon slopes mostly that ferns love to grow; in places where water rarely lodges, but where moisture is abundant, and there is some shade against the noon-day summer sun. Note all you see of the whereabouts and ways of your favorites, and you will find that there is a better book on fern-growing than the one you are now reading—it is the Book of Nature.

CHAPTER III.

HOW TO FORM AN OUTDOOR FERNERY.

TO keep up your interest in the subject, make a fernery at the very outset, even if you do not know the names of half a dozen ferns. If you cannot go collecting you may be able to dip into the tempting basket of the itinerant fern vendor, who is sure to be able to supply you with the Male fern, or *Lastrea Filix mas,* which is the hardiest of all, and will grow almost anywhere; the Lady fern, or *Athyrium Filix fœmina;* the Hard fern, or *Blechnum spicant;* and the Hartstongue fern, or *Scolopendrium vulgare.* With these four you can make a good beginning. It is usual to construct the outdoor fernery of some sort of "rockwork," and for two good reasons, first, because the forms and hues of ferns are more effectually displayed when their bright green tufts rise out of grey stones or dark burrs from the brick kiln; second, because they thrive better, when planted in gardens, if their roots are protected from excessive evaporation by the covering of the soil with stones and vitreous masses. Many a tiny fernery do I see in my travels placed at the entrances to country villas and cottages, where I should never think of placing them, yet they look quiet and pleasing, and suggest to all passers by that those who planted them did their best to vindicate the quiet beauties of God's great harvest, knowing that for more demonstrative forms of vegetable splendour vindication was unnecessary. When little ferneries like these are constructed, only the commonest and most robust-growing ferns should be planted in them. Generally speaking, the common soil of the place will do, but if a quantity of leaf mould or cocoa-nut fibre can be mixed with it the better. If there is any doubt about the soil of the place being suitable, get some sandy or peaty earth from a common where ferns and heather are found in plenty, and have enough to raise the position above the general level, then cover it with stones or burrs, and plant the ferns between. There are sorts well adapted for this simplest form of fernery, namely, the four just named, as likely to be found in the fern dealer's basket, and the following:—the Bracken or Brake, *Pteris aquilina,* the Broad Prickly Buckler fern, *Lastrea dilatata,* the royal Osmund, *Osmunda regalis,* the common Polypody, *Polypodium vulgare,* the Common Shield fern, *Polystichum aculeatum.* Many more may be added if the soil is a mellow, friable yellow loam, with plenty of sand in it, but it will be well to get a little used to ferns before launching out into grand speculations. When you have had some practice in this humble way, and have, perhaps, succeeded in growing a few ferns in pots in a frame or in a fern case in the drawing-room, you will become ambitious, and resolve on having a grand fernery, with, perhaps, a model of a ruin for the main feature of the scheme.

Outdoor ferneries are usually formed of tree roots and banks of earth, picturesquely disposed and planted with ferns severally adapted to the sites and positions the scheme affords. Where there are living trees on or near the spot (and the shade of large trees is desirable), the use of roots is objectionable, because of the quantities of fungi which are sure to be produced, the mycelium from which may find its way among the living roots and commit vast havoc. But even this danger is worth risking sometimes in cases where roots and butts are plentiful on the spot, and it is undesirable to incur any great expense. The foundation of all banks and earth-works for ferns should be good loam or clay, into which many of the stronger-growing kinds will send their roots when well established. But the upper crust and the stuff for filling in between roots, burrs, &c., should consist of half peat and half silky yellow loam, or some mixture which nearly approximates in character to such a combination. Thus, good loam with well-rotted cocoa-nut fibre, or loam mixed with yellow leaf-mould and manure that has lain by three or four years till rotted to powder. It is best to complete the structure and fill in all the more important places intended for soil before inserting any of the plants, for the simple reason that the work must be firm, the soil well rammed in, and the whole of the scheme so substantial that there will be no fear of any portion shrinking away afterwards, and leaving the roots of the ferns without soil, or causing hollows and crevices between the blocks and the banks into which they are set.

ROCKWORK AND COMMON BRACKEN.

My own outdoor fernery was figured and described in the 'Floral World' of January, 1867. It consists of walls and arches forming a sort of ruined bastion. It is entirely built of "burrs" from the brick kiln, which is the best material for the purpose in districts where rough stone is not to be obtained. All the walls are double, and filled in with strong loam, and, of course, are roughly built, with many crevices and hollows, in which the ferns are planted. These walls may be likened to cases containing earth which is fully exposed on the summit to the weather, and consequently may be regarded as another kind of banks. The annexed diagram will give an idea of the *principle* of construction, though straight lines of course convey no idea of their form.

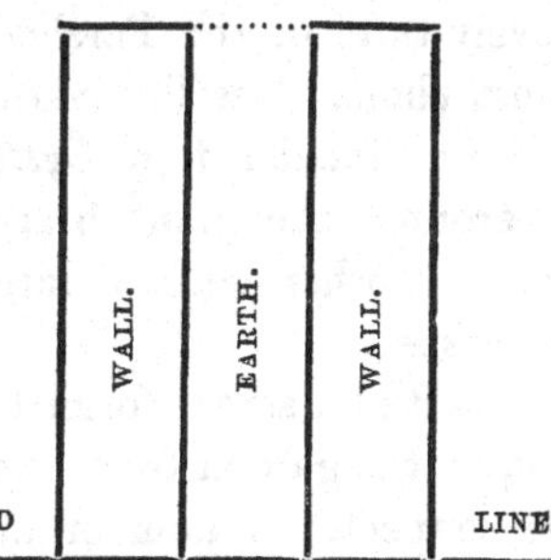

Where the walk passes through the bastion, the walls rise clear out of the gravel, but all round in the bays and inlets mounds of earth are raised against them, as would be the case in a real ruin, from the accumulation of rubbish. As a hint of the rough construction of the walls, and the nature of the effects produced, here is a "bit" of scenery from the bastion, from a "photo,"

showing how the bracken grows on the rubbish heaps in nooks amongst the walls. The whole scheme is planted with ferns, and various flowering Alpine and rock plants, every position having forms of vegetation suited to it. Thus, at the base, where the walk passes through, there are great tufts of lastrea and lady fern; on the summit, crowning the work, and rooting into the great mass of earth, the common polypody thrives as bravely as on the pollard alders and oaks in Epping Forest. High up in dry positions, on the face of the wall, grow the Wall Rue, *Asplenium ruta-muraria,* the Maidenhair spleenwort, *Asplenium trichomanes,* with many varieties of sempervivum, sedum, thyme, and other plants that love such positions. On the smaller knolls, and in half-shaded bays, where there is a good depth of earth, may be seen lovely tufts of the Parsley fern, *Allosorus crispus,* the most choice tasselled varieties of Harts-tongue, the delicate Bladder fern, *Cystopteris fragilis.* On the banks around, the giant bracken towers up above our heads, and other ferns of large growth congregate in rich masses.

My bastion is part of a screen formed to separate the pleasure division of the garden from the experimental, and with it are connected a number of features, such as a rustic house used as a summer reading-room, a bee-house, some great tree butts planted with ferns, ivies, and grasses. I am satisfied that where space can be afforded the imitation of a ruin is the best possible central idea out of which to develop a fernery.

We shall have to refer to rockeries again in various ways, but as I am resolved to make no long, tedious chapters if I can help it, I will here offer a few general advices on the formation of ferneries out of doors.

Provide as many aspects and degrees of declivity as possible within certain limits. One slightly irregular bank is to be preferred to a number of paltry ins and outs, but if you have space and materials sufficient, let the work be somewhat intricate in order to obtain a *variety of conditions* to suit the various habits of the ferns you intend to grow.

Large bodies of soil are absolutely necessary, as it is impossible to keep the roots moist enough during the hottest months of the year if they are in shallow soil, of which a large surface is exposed to the atmosphere. It is particularly important to bear this in mind in constructing the walls of a ruin, if it is intended to plant ferns on or in the walls. A space of one foot clear, filled in with earth, between the two faces of the wall, is the least that should be allowed in the smallest construction of the kind; two or three feet of earth will be required in a ruin of dimensions large enough to serve as a garden-house or reading-room.

Aim at wildness and apparent neglect in the arrangements up to a certain point. Dirt and disorder are as injurious to the ferns as to the morals of those who encourage such things, but primness is not desirable in a fernery; the effects should tend towards the *rustic* rather than to the refined, and the materials used throughout should be of the quietest colours; no gew gaws, no plaster casts, no blocks of coral or shiny shells should be mixed up with the work.

Robust-growing ferns planted on banks and mounds

of good mellow loam will scarcely want cultivating. Pretty well the best you can do for them is to leave them alone. But those elevated on pinnacles and in other positions where they are likely to get very dry must have the aid of water, not only in *summer* but in *winter*, also on every occasion when dry weather prevails for any length of time. Many plants so situated perish by desiccation during the prevalence of east winds in March, when because the weather is cold and they are not growing, the cultivator is apt to think water unnecessary; or rather he is apt not to think about the ferns or the water at all.

Small-growing delicate habited ferns that are in exposed positions on rockeries should have protection during severe frost. A flower-pot may be inverted over them or a little clean hay may be placed over their crowns and kept from blowing away by means of a few pegs, or cocoa-nut fibre or sand may be heaped up round and over them, to be taken away of course when the crowns begin to throw up new fronds in spring. Always wait for mild moist weather to remove such protection, for if the swelling crown is suddenly exposed to a cutting east wind, it may shrivel and perish, instead of throwing up its emerald tassels in token of the return of the tender spring.

Thus far we have considered outdoor ferneries as superstructures. We might have regarded them as substructures. At all events, I should like for an old quarry to become mine some day that I might make a fernery of it; and perhaps lacking a quarry, I may be tempted to throw myself into a gravel pit, and by a little hard work and patience make a fern garden of it.

CHAPTER IV.

CULTIVATION OF ROCK FERNS.

YOU have taken notice when fern collecting that many of the smaller kinds are only found on rocks and old walls, or, at all events, are never found in damp hollows or in places over much sheltered from the sun and the breeze. Now, all such ferns require peculiar treatment, and as you advance in practice the rock and wall-loving varieties will probably interest you more than all the rest.

The first requisite to success is to plant them where it is impossible for water to become stagnant about their roots. In planting them on a rockery it is a good plan to take out a quantity of the soil from the place where the fern is to be, and introduce soil specially prepared for it.

In preparing the stations put a lot of broken bricks or broken flower-pots and small stones into the holes, and upon these let there be full nine inches depth of the compost, and let it be raised into a hillock.

Nearly all the ferns of this class will thrive in a mixture of equal parts of yellow loam of a silky nature, fibrous peat or the top crust of sandy soil from a common where the ling and the brake grow together. There must be full one fourth of sand in the mixture, but the loam or peat may contain as much as that, and no more need be added. If the loam and the peat are both of an unctuous nature, add sharp sand in quantities equal to one fourth or even one third of the bulk, and mix all well together. *Never use sifted soil* for ferns (except in the case of seedlings, to be spoken of in a future chapter), but have all lumps broken to the size of walnuts or hazel nuts, and mix fine and coarse together.

In planting the ferns, those that have a creeping rhizome or root stock must be slightly covered, and it may be necessary to fix them in their places with a few pegs. Do not cover them deeply, only so much in fact as to prevent exhaustion of the rhizomes by drying winds until they can make fresh roots, by which time the frequent sprinklings they are subjected to will have washed the mulching off the rhizomes, which will then be left in their natural position *on* and *not in* the soil.

It will be well perhaps to make a few remarks on the species which come into this group. *Allosorus crispus*, the mountain parsley fern, makes a charming tuft on a rockery; it is fond of stone, and abhors damp. I find that a mixture of equal parts peat, decayed cocoa-nut fibre, and broken pots or broken hearth-stone suits admirably. It must be shaded, or the new growth soon goes rusty.

Asplenium adiantum nigrum, the black maidenhair spleenwort, is rarely met with but in positions elevated above the ground; it greatly needs shade and shelter, and will thrive in any peaty mixture, or in broken pots alone.

Asplenium ruta muraria, the wall rue, requires a very

ASPLENIUM VIRIDE.

dry and open position, and will do well in a mixture of two thirds broken bricks and chalk, and one third sandy peat. Stagnant moisture will be speedy death to this fern, but it must have daily sprinklings while growing to promote free growth.

A. septentrionale, the forked spleenwort, should always be grown in an elevated position for the sake of the protection thereby afforded it against slugs and wood-lice, which rarely get into the higher parts of mural ferneries. Being very small, it may be easily lost when planted on banks or level ground; but in a suitable pocket in a sheltered nook in a wall or ruin, it makes a very pretty and interesting patch.

Asplenium trichomanes, the common maidenhair spleenwort, and *A. virides,* the green spleenwort, are superb wall ferns, and in fact they rarely do well under cultivation except when planted out in an elevated and well-drained position. The soil should be equal parts sandy peat, yellow loam, and broken bricks, and the plants should be planted firmly, with their crowns slightly above the surface.

Ceterach officinarum, the scale fern, is essentially a wall or rock fern, and a very beautiful and interesting species. Confinement and damp are most prejudicial to this fern, and when planted on a rockery under glass the most airy position safe against drip should be chosen. Any good sandy soil will suit it.

Cystopteris montana, the mountain bladder fern, requires peculiar care. Select for it a position thoroughly sheltered and shaded, and prepare for it a station with a stratum of broken bricks for drainage, and over that six inches of a mixture consisting of sandy peat, sphagnum, and broken sandstone or common hearthstone. Plant in the centre of the station, and place a bell-glass over; keep constantly moist, and give air periodically. When it is well established, remove the glass, and leave it to take care of itself. If the fernery is supplied with a stream of water, Cystopteris montana is one of those which should be planted on a ledge of rock where it can have the benefit of a daily trickling of water over its rhizomes.

Lastrea montana—the hay-scented fern, better known, perhaps, as *L. oreopteris*—requires similar treatment to that recommended for Cystopteris montana, but should have a soil more inclining to loam. It can scarcely have too much water, provided the position in which it is planted admits of it readily flowing away.

Polypodium vulgare, the common polypody, will grow in almost any position except in a sheer marsh, and there it soon perishes. When growing wild in the woods, whether on pollard trees or moist banks, it is invariably found rioting in deposits of leaf-mould and wood rotted to powder. Pure cocoa-nut fibre, or equal parts of the fibre and mellow loam, pure leaf-mould, and very dry, tough, fibry peat, in which there are old hummocks of grass, are soils that suit this fine fern to perfection. It will bear sunshine well, but grows more luxuriantly in the shade. In a very dry position where no water can lodge about it, but sprinkled daily all the summer, this fern will attain to grand dimensions, and be one of the most beautiful in the collection all through the autumn and winter months.

Polypodium Robertianum, the limestone polypody, requires a dry position, and a mixture of sandy loam and chalk.

Woodsia ilvensis, the oblong Woodsia, must have shade and shelter: and the most perfect drainage. Make a little hollow of broken bricks, or other porous substances. Fill with a mixture of yellow loam and silver sand. In this the plant will luxuriate.

GROUP OF SCOLOPENDRIUMS ON ROCK-WORK.

CHAPTER V.

CULTIVATION OF MARSH FERNS.

THIS will be a very short chapter, just because there are no marsh ferns. I remarked as much in Chapter III, and pointed out that the most moisture-loving of them managed usually to keep out of the water. But you may wish to plant some ferns beside a stream, or on an islet, or near a fountain, or in some other peculiarly damp position, and it will be proper here to name the most suitable.

Osmunda regalis, the royal fern, delights in moisture, especially if it is growing in a great bed of spongy peat. With such aids and a warm climate it will overtop the tallest man, but if it only attains a height of five feet, it is a noble object, as much like a palm as any plant of English growth.

Athyrium Filix fœmina, the lady fern, delights in a similar position. This has no palm-like aspect, but is rather to be compared with a plume of ostrich feathers of the most intense and delicate tint of yellowish green.

Lastrea thelypteris, the female buckler fern, is another charming species for a very damp position, and it spreads fast, literally carpeting the ground with pale greyish-green most delicately textured fronds.

Blechnum spicant, the hard fern, will attain grand

dimensions, and produce abundance of fruitful fronds in damp spongy peat. I never saw this and the Osmunda grow so grandly as in a wet gully I struck upon once when fern hunting in a wood near Oakshot in Surrey. There the Osmund was my equal in stature, and the fruiting fronds of the blechnum just reached my chin. It was a very damp spongy spot, yet the ferns stood a little above the water line.

Charming plants to associate with the moisture-loving ferns are the *Equisetums* or Horsetails. Get *Equisetum sylvaticum* if you can, and plant it in wet spongy peat in a sheltered nook, and you will have a bit of vegetation that will make you proud of the land of which it is a native,—that is, if you happen to be a true Britisher, which the plant is,—if not, be glad now and then that you came here, for if this is not in any especial manner the land of ferns, it is at all events the land of people who love them.

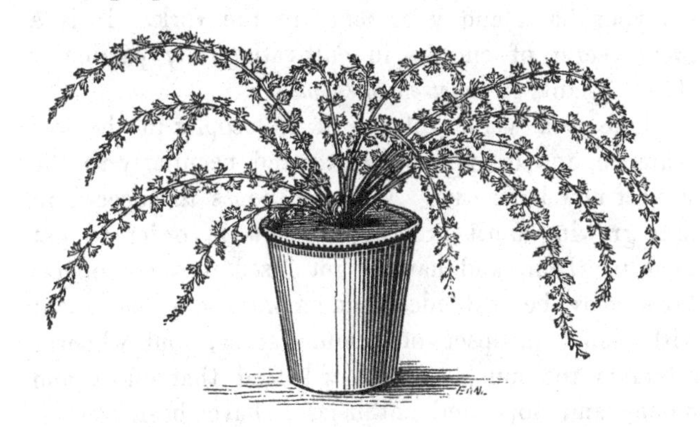

ATHYRIUM FELIX FŒMINA, VAR. FRIZELLIÆ.

CHAPTER VI.

FERNS IN POTS.

LET us now make another advance in practice. Ferns are beautiful objects when well grown as pot plants. To grow them well in pots demands more care and skill than growing them in the rockery, because there they, for the most part, take care of themselves. But pot plants are at all times more dependent on the cultivator, and must have constant attention. If you fail at first do not be discouraged, for the practice is attended with but few difficulties. Begin with a few of the commonest, and do not make a rush at rare varieties, until you have got your hand and your *mind* in the work. It is a great secret of success in cultivating any particular class of plants *to get used to them.*

There is a whole volume of philosophy in the last sentence, and it applies directly and peculiarly to the subject now before us. Whoever hopes to succeed in fern growing must first grow a few in order to get used to them, and having got used to them operations may be extended and money may be spent with some prospect of remuneration; but whoever attempts too much at first will find that effort and money and hope and enthusiasm have been wasted, for disappointments in the early stages of a pursuit are ten times more dispiriting than when they occur after we have been rewarded with many successes.

Suppose the beginner in fern growing to take in hand a dozen species only; what shall they be? I should recommend the following:—*Lastrea Filix mas, Lastrea dilatata, Polystichum aculeatum, Polystichum angulare, Polypodium vulgare, Polypodium dryopteris, Athyrium Filix fœmina, Asplenium marinum, Scolopendrium vulgare, Cyrtomium falcatum, Woodwardia radicans, Lomaria chiliensis.* These are among the cheapest and most easily procured. The first eight of them are British, and the remaining four foreign. Supposing them to be all small nursery plants, they might all be potted in five-inch pots, or what are termed 48's, but the size of the pot must depend upon the size of the plants, and that size will suit which will take their roots without cramping them, and allow very little space beyond. Fine specimens can be grown with more certainty by shifting them into larger and larger pots as the plants increase in size, beginning with pots as small as possible without cramping the roots, than by putting them into large pots in the first instance. The soil that would suit all these would be such a mixture as the following: one part peat, consisting of the top slice of turf, and which consists chiefly of the fibre of fine grasses, the roots of heaths, decayed moss, &c. This must be chopped up the size of walnuts. One part friable yellow loam of a clean *silky* texture, such as will crumble to powder between the fingers, and yet scarcely soil them even when it is moderately damp. If this is full of fibre of grass turf, and has the fleshy roots of brake intermingled with it, all the better. Such loam as this is very abundant, indeed it generally prevails where the brake grows luxuriantly in the hedgerows. One part thoroughly decayed leaf mould, which should be black and gritty, free from fungus, and from bits of iron and other rubbish which gardeners too often allow to get mixed up with it. One part silver-sand. Mix these ingredients well together; break all lumps to the size of walnuts; do not sift it, and do not on any account endeavour to make it fine like dust. Indeed, *a compost as fine as dust will not grow any plant to perfection.* When prepared, the sand should be visible throughout the mass, giving it a grey hue and a granular appearance. It should be only moderately moist, *not wet,* free from large stones, and have a pleasant feel in the hand. Now draw to one side a heap of the toughest and largest pieces of fibre and loam from the mixture; this we shall call *rough stuff.* Next lay ready for use a small heap of green moss, or, if not green, tolerably tough and fresh, that is to say, not rotten. Next break up a lot of flower-pots to the size of crown-pieces, and another lot to the size of peas. The last job preparatory to potting is to have new or quite clean pots. If they are not clean inside and out, the ferns will not thrive, and if they did we should not like them for it.

The process of potting is very simple, yet it is usually badly done by beginners. First place over the hole in the pot a picked potsherd, hollow side downwards; then lay, also hollow side downwards, a few pieces all round, to cover the bottom of the pot, and then add a good handful of the smallest potsherds. Next spread a thin layer of moss, then a thin layer of the rough stuff, and then take the fern in the left hand and place it with the crown in the centre, level with the rim of the pot, and allow the roots to spread, so that when earth is put upon them they will not be cramped up in a bunch. Take the compost in the right hand, and pour it in all round till the pot is full, and then with the thumb of each hand press it down, turning the pot round in so doing, adding more soil as required, so that when finished there will be half an inch of space between the soil and the top edge of the pot. There is a golden rule for success in growing any kind of plant in a pot, and it is *to pot firm.* Do not be afraid to press the earth in round the roots, and give the pot a tap on the board at the finishing touch; when potted loosely, no plant can thrive.

Now, what are we to do with this dozen of ferns? I propose that we fit up a frame to face the north in some quiet corner of the garden, and that we make no boast about our ferns until they have had one year's growing at least. We want a dry spot, rather sheltered; the soil on which the frame is to stand should be covered with coal ashes, and be easy of access at all seasons. Suppose we have potted them from the 30th of March to the 1st of May—ferns may be potted at any time, but when they are just starting into new growth is the best time—the next question is, What shall we do with them? It is but little they require; first place them in the frame, next water them with a common watering-pot with a fine rose on the spout. When you get used to ferns, you may water them without the rose, unless you wish to wet the fronds, but you must use the rose now, because, as you are not used to them, you might wash half the earth out of the pots by a sudden dash of water, a contingency not possible when the rose is used in watering. All through the summer these plants will want a little water every other day at least, and it should be given so as to wet the fronds all over, and moisten the soil *without drenching them.* In very hot and very dry weather daily watering will be necessary, and in the very hottest weather you may water twice a day with benefit.

Not less important is the giving of air and light. If the frame faces north, the light may be stood up on end, leaning against the back, so as to form a sort of south wall to the plants, and a mat hung on it, or a breadth of canvas tacked to it will render it efficient to screen off the full blaze of sunshine. If this cannot be done, put the light in its place, lay a mat upon it and draw it down, and tilt it slightly with blocks of wood or empty flower-pots, so as to allow a current of air to pass through. In this state it is to remain from the 1st of May to the 1st of September, *during the day time only.* Every evening—at sunset or earlier—draw the light off altogether, that the plants may have the full daylight as long as it lasts, the dews all night, and the full daylight again in the morning till about 9 a.m.

On the 1st of September your plants will have a most luxuriant appearance, and the pots will be crammed full of roots. Shift them all into pots the next size larger without breaking their balls of roots, and let the operation be performed in precisely the same manner as already described. Take off the shading, and give the plants very much air both day and night for another month. During very bright sunshine shade them for an hour or two; but let them have the sunshine morning and evening, and the night dew. Continue to water as before, but give less and less, so that by the 1st of October they will be watered only once a week. After that date, until frost occurs, shut them up at night, take off the light all day, and once a week pour water gently over their crowns, sufficient to make the crowns moist, but not to sodden the soil in the pots. When frost occurs, throw a few mats on the light; if the frost increases in severity, take off the light, and strew *dry hay* amongst the plants, taking particular care to cover their crowns with this protecting material. Put the lights on, lay some dry straw or hay on the glass, and then lay a good mat over all. If you neglect these precautions, you will probably not lose any of your plants, for they are all hardy; but the effect of frost upon them will be that they will be a week or two later in growing in spring than if they had been protected, and so I must insist on protection as necessary.

Whenever the weather is mild, or the frost only amounts to a few degrees at night, continue to pour a little water over the crowns once a week; in fact, that operation is only to be suspended when the weather is really severe. Give air as often as possible, never allow them to become thoroughly wet, and keep them safe from being frozen.

The majority of amateur fern growers allow their pot plants to go dry as dust all winter, and the consequence is that they grow very poorly in the early part of the following season; in fact, scarcely grow at all till June, by which time their new fronds ought to be all completed. It is a grand secret of success to keep their crowns freely moistened all the winter long.

The next best time to shift them will be the 1st of March. Proceed as before, using pots one size larger. You will now have fine specimens. The frame will no longer hold them. You must either build a greenhouse to keep them in, or you must have a pit of sufficient depth to give them head room, or you must make a rockery and plant them all out in it, or you must divide them all by splitting them asunder with a knife right through the crown, and pot all the pieces, or you must sell them and retire on the proceeds. It cannot be my business what becomes of them after this date; it suffices that I have made a fern grower of you, and you will be enabled to understand and practise all the directions and suggestions on fern growing which you may find in this volume or any other that may be worth referring to. You will have learnt that a clean, granular, peaty, fibrous soil; a rather still, warm, and moist atmosphere, and shade from sunshine, are the principal essentials to success in fern growing, and to make short of this part of the paper, I may as well say that you have very little more to learn in the way of principles; if you are ever to excel in fern growing, it will be owing to the use you make of observation and experience in carrying those principles into effect.

CHAPTER VII.

THE FERN HOUSE.

WE are now becoming " expensive and hard to please." We want a fern house—oh dear! how our wants increase with increase of knowledge and advance of taste. Any man could live contented on just double the amount of income he has already, and the fern grower at any time could promise to be satisfied if he could be sure of advancing from a frame to a house, or from a house to another and a larger house, and from such ferns as anybody could grow in a modest cool fernery to tree ferns of gigantic growth, and the gorgeous *Leptopteris superba,* which is perhaps the loveliest fern in the world, and rather too dear as yet, and needing too much care for the humble fern grower ever to dream about it.

By a fern house I mean some sort of cave or rockery covered with glass and with or without heating apparatus. The best example of a fernery of this sort I know of, to which the public have access, may be seen at Messrs. Veitch and Sons' nursery, Chelsea. It is truly a garden with gravel walks amidst rocks and waterfalls, and on every hand the ferns present themselves in sheets of delicious verdure or in waving palm-like masses, or in a glorious confusion of brake and lastrea intermingled as if the dryads themselves attended to the planting. I could mention hundreds of private gardeners where I have seen beautiful ferneries under glass, but the reader would gain nothing by the list. Pardon my boldness, but in truth I have scarcely met with a fernery to surpass Mrs. Hibberd's in beauty and interest, though it is on an extremely small scale. I will tell you something about it.

Given, a recess in the walls of a house, and what shall we do with it? It is of no use to put the question to echo, who is represented as giving answers as required, because an honest echo could only reply, "Do with it!" which, at the best, would be ambiguous, and might be supposed to mean, "Do *away* with it!" In a certain sense that is just what I have done; for, by converting the recess into a fernery, it is a recess no more, but a part and parcel of the garden, and yet not utterly separated from the dwelling-house. Please allow a few hap-hazard lines to represent the case in the first instance. If you suppose A to be one side of the house,

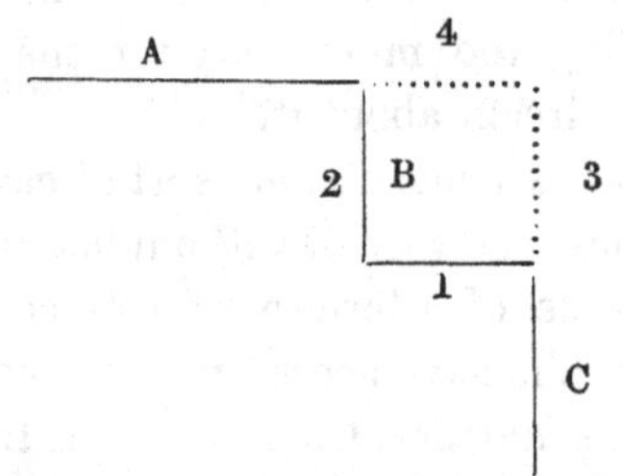

and C another side, then B will be the recess or hole in the wall requiring to be occupied in some way or other, or by some construction to be blotted out. Letter A looks west, letter C north; the garden-walk passes by the side of the house along the line A, and past the gap B; and as long as that remains a gap, it is abominably ugly. It is twelve years ago since I filled up the gap B with a lean-to greenhouse, with the slope of the roof looking west, and the door on the side which looks north. Fig. 1 is the back wall of the house, fig. 2 the

end wall, fig. 3 the door, fig. 4 the front. This was at first used as a small show-house, for, being easy of access, always in sight, and in a shady position, it served the double purpose of displaying a few good things in a place where it was convenient to see them, and also, by reason of its cool, shady position, keeping them longer in perfection than they would have remained in any more sunny position. In the cut the house is shown with a stage for flowering plants, as originally constructed. In the course of time, some building and planting took place a little way off towards the west, and the nice gleam of sunlight that enlivened the house from 2 p.m. till sunset was effectually blocked out, and the house became unfit for flowering plants. Instead of bringing an action against the neighbour who devoured my sunshine, I brought an action against myself, and the verdict was, that the shady house should be forthwith converted into a fernery. The stages were removed, and in their place a rockery was built upon a very simple plan, and which, considering the smallness of the house, proves delightfully effective, as affording at all seasons a beautiful scene, and very serviceable arrangements for the growth of plants. I employed a skilful bricklayer to do all the solid work, and, under my direction, he faced the back and end walls of the house (1 and 2) with a rugged mass of burrs from the brick-field, rendering it somewhat like the interior of a cave. The work was commenced at some little distance from the wall, and gradually brought nearer and nearer as it proceeded upwards, occasional large blocks being firmly cemented to the wall, and strengthened with holdfasts; and between the walls and the burrs good loam was rammed in from bottom to top. Next the front wall (4) and the end (3) a low border was formed with a facing of burrs, this border consisting of good loam. No special device for drainage was resorted to, and it has never been wanted; a layer of broken bricks, about six inches deep, was put upon the tiles, and the soil thrown upon this rough bed. There is a trapped sink leading to a drain in one corner of the house, and all superfluous water finds its way there quickly, as the pavement slopes gently to it. The finish of the work I did myself, and it occupied me, at odd times, about four months, the work being essentially amusing, though attended with an occasional abrasion of the knuckles. The task I had was to make the "pockets"—openings for the purpose being left at intervals in the work. I made the "pockets" and planted the ferns at the same time. Some of the larger ones are planted in projecting receptacles, just as the bricklayer left them; but generally speaking, I found it the best plan to stuff the necessary soil into a chink or gap, then place the fern in it, and, lastly, to introduce a piece of burr of suitable size to close it in, and this was done with the help of cement. I do not think I can profitably occupy further space with remarks on the formative part of the affair; details of this kind do not admit of being described minutely; all I can say in concluding this part of the history is this, that I never did a better job in my life; for not only have the ferns and mosses planted in it thriven amazingly, but the scene produced is exquisitely beautiful and affords more than a suggestion of the

"Negligence of nature, wild and wide."

It is of the utmost importance to inform the reader that the house is not heated. It is remarkably proof against frost, which I attribute to the fact that the back wall (1) forms one side of the drawing-room, which is kept at a comfortable temperature all the winter, and of course the wall itself is in winter always warmer than the atmosphere outside the house. That frost does get in, however, is certain; the thermometer several times indicated five to ten degrees of frost inside, and when the case has become in any way serious, Hays's constant stove or Hinks's petroleum stove has been set to work to keep all safe until the weather changed for the better.

A fern house is a genuine luxury, which every lover of ferns should have if possible. Amongst its many advantages, a few must be named as particularly worthy of attention. It affords, even without the aid of artificial heat, opportunity for growing a number of nearly hardy ferns which need some protection, yet are not much hurt if they have to endure a few degrees of frost. Ferns of this class are numerous and extremely beautiful. I will name four only now as a key to the rest in illustration of this particular advantage—*Woodwardia radicans, Cyrtomium falcatum, Adiantum pedatum, Todea pellucida.* Another advantage is that if planted only with the hardiest British ferns, they grow more beautifully than the same sorts do in the open air. As a winter garden and as a peculiarly charming scene —if well done and well kept—the fern house is worth something to a home bird, and as an amusement for an invalid it is invaluable.

The management is a modification of that advised for the outdoor fernery, but watering must be more regularly performed, and if fast growing ferns run riot and overrun the others they must be kept in check by occasionally digging out their roots. Lastrea thelypteris and Onoclea sensibilis are likely to do this, but they are such lovely inmates of a cool fernery that you cannot do without them.

During the summer the roof must be shaded with thin "scrim" or "tiffany," or a smear of whitewash. At all events, it will never do to allow powerful sunshine to shrivel up the tender growth and change the glistening green to dingy brown, as it will do very soon if there is no shading used.

A very small amount of ventilation will be required if the house faces north, as it should do. A fern house in a hot southern exposure would need abundance of air, heavy shading and extra help from the garden engine from the 1st of May to the 19th of September. After the last-mentioned date it might be left alone for ever, for one season's struggle against overwhelming odds ought to be enough for anybody. Choose or make a shady place for your house, and then see that, as the rockery is built up, there is a good body of earth for the ferns to root into.

Constant attention will be requisite to keep the fern-house as beautiful as it should be. Dead fronds must be removed without injury to the young fronds that are rising; some ferns will need more water than others, and in the height of summer the floor must be wetted daily to cause humidity of the atmosphere.

The following ferns have prospered in the house during the past twelve years, all of them having been at times exposed to a few (say half a dozen) degrees of frost. It must be borne in mind that they can better endure frost if planted out than in pots.

LIST OF FERNS FOR PLANTING OUT IN A GREENHOUSE FERNERY:

LARGE GROWING FERNS.—*Woodwardia radicans* and *W. orientalis*. Plant these about five feet above the ground, that their drooping fronds may be seen to advantage. *Cyrtomium falcatum*, a rigid grower, often and appropriately called "the laurel fern." *Onoclea sensibilis*, suitable for a cool damp nook on the ground line, as it is an upright grower; under glass it is a magnificent fern. *Lomaria chilense*, a rigid habited fern with bold dark green fronds. *Pteris flabellata*, grand pectinate fronds, of a vivid light green colour, suitable for a shelf or bank three or four feet above the ground. *Pteris cretica albo-lineata*, an upright grower, spreads freely at bottom. *Phlebodium sporodocarpum* requires a well-drained elevated ledge, from whence it will put forth masses of tawny roots and handsome glaucous fronds. *Polystichum acrostichoides*, a bold habit, and a good companion to *Lomaria chilense*. *Lomaria magellanica, Lastrea intermedia, L. frondosa, L. Sieboldii, Asplenium angustifolium, Adiantum pedatum, Athyrium asplenioides, A. tenuifrons, A. filix-fœmina v. corymbiferum.*

MEDIUM GROWERS.—*Davallia canariense, Asplenium Michauxii, Adiantum cuneatum, A. assimile, A. affine, A. formosum, Asplenium bulbiferum, A. angustifolium, A. athyrium filix fœmina v. Frizelliæ, A. f. f. v. corymbiferum, A. f. f. v. crispum, Scolopendrium vulgare v. crispum, S. v. v. alcicorne, S. v. v. ramosum, S. v. v. ramo-cristatum* (and a dozen others of the same series desirable), *Lastrea thelypteris, L. æmula, L. Goldieana, L. filix-mas v. cristata, Polypodium dryopteris, P. aureum,*

P. phegopteris, Platycerium alcicorne (suitable to suspend on a block of wood; it is almost hardy). *Todea pellucida* (this grows finely in a cool house, if in a damp, shady, and *still* place; wind it cannot endure).

For Elevated Positions in the Fronts of Rockeries—that is to say, to grow as wall ferns, and all requiring plenty of air: *Asplenium trichomanes, A. adiantum nigrum, A. marinum, Ceterach officinarum, Allosorus crispus* (a sunny position near the door will suit this and Asplenium trichomanes), *Polypodium vulgare* and its varieties, especially *Cambricum* and *Hibernica.*

The most hardy of the tree ferns is *Dicksonia antarctica,* which is as easy to grow as a common lastrea, provided it has enough water.

For Baskets, take *Pteris scaberula, Adiantum setulosum, Asplenium flabellifolium, Camptosorus rhizophyllus, Davallia pyxidata, Niphobolus lingua,* and any of the free-growing hardy ferns that run about freely, such as *Lastrea thelypteris* and *Onoclea sensibilis.*

If you should wish to create in your house or out of doors a constant trickling of water for the benefit of some fountain-loving ferns or mosses, take any large vessel, in the bottom of which you can break or bore a small hole. Cover the hole with a flat tile, and over that put two inches of the finest sand. Fill the vessel with water daily, and it will run gently as long as there is a drop left at the fountain head. If an ornamental vase should be used for the purpose, it might be utilized by placing in it a pot containing some semi-aquatic plant.

CHAPTER VIII.

THE FERNERY AT THE FIRESIDE.

THOUSANDS of amateur fern growers have only a glass case in the sitting-room for a fern garden. In the heart of a great city where gardens are unknown, and even the graveyards are desecrated by accumulations of filth, the fern case is a boon of priceless value. It is a bit of the woodside sealed down with the life of the wood in it, and when unsealed for a moment it gives forth an odour that might delude us into the belief that we had been suddenly wafted to some bosky dell where the "nodding violet grows." Before we go a step further it is but just to the memory of a good man to call to mind that for many years the structures now commonly called "fern cases" were known as "Wardian cases," being the invention of the late Mr. B. N. Ward, an eminent surgeon, many years resident in Finsbury Circus, who died at a ripe age in 1868. Peace to his memory! He not only added to the embellishments of the English home and the recreations of English domestic life, but his invention has been of incalculable service in the introduction of valuable exotic plants to this country, for if shut up close in Wardian cases they travel over sea far more safely than by any other system of protection.

The simplest form of a fern case is the bell-glass and flower-pot, of which the annexed sectional figure affords an accurate representation. This particular form of pot was invented by Mr. Fry, of Lee, and is made by

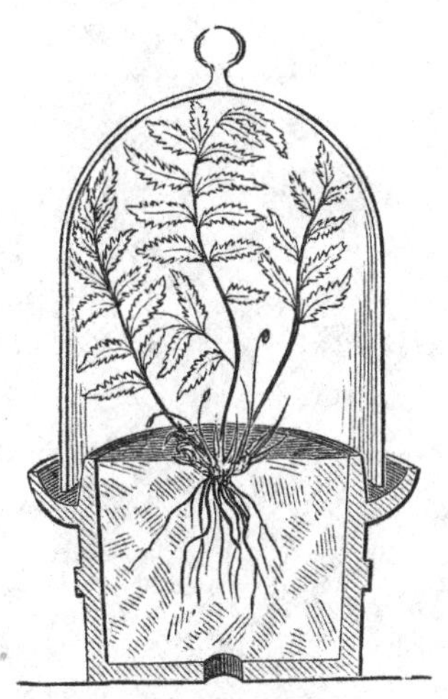

Mr. Pascall, a potter at Chiselhurst. It can be obtained of the dealers in ferns, and forms a very neat table ornament, as the pot is made of fine red ware and is roughly ornamented.

Another form of the same kind of tning consists of a glass dish with rim and bell-glass, the whole very neatly finished, and forming, if skilfully fitted, a most elegant miniature fern garden.

Fern cases constructed of wood or metal frames, with boxes or troughs for soil, have been made in endless variety, yet for real utility and beauty of appearance there are none so good as those of the simplest rectangular outlines, such as may be readily obtained of any of the leading dealers in such things. All things considered, the cases known as "Miss Maling's," which

may be heated if required by means of gas flame or hot water renewed periodically, are the best, because of their extreme simplicity and the uninterrupted view they allow of the interior. We are supposed to be great in fern cases—I say *we* in the way of the organ blower in the story; but Sine qua Non is the master of them here; and they comprise cases of several kinds, some rather gigantic in dimensions, besides vases fitted with lanterns of plate glass, in which not only climbing ferns and ferns of many other kinds, but climbing ivies, lycopodiums, and such odd things as the artillery plant are grown. I have had to make as well as furnish plant cases. We have between us managed to exhibit a considerable number, and step by step we have acquired some very definite ideas about them, which I shall endeavour to set forth categorically.

In the case of fern shades which fit into glass dishes, and which, as long as there is water lodged in the rim into which the lower edge of the shade rests, are air-tight, air must be given three times a week by removing the shade altogether for an hour or so. This allows the excess of moisture to dry off the foliage, and prevents mould; and the glass getting dry in the meanwhile, it is prepared to take up a fresh supply of moisture from the soil when replaced, which is equivalent to a circulation of water as well as a change of air. This air-giving, however, must be regulated by discretion, for if the air of the room is hot and dry sudden exposure of the plants to it may do them harm. Moreover, it is a very easy matter to remove the glass and *forget* it, the result being, perhaps, complete destruction of all the more tender fronds, and the disfigurement of the affair for a fortnight. Now, a very simple and expeditious and effectual mode of ventilating consists in taking off the glass, wiping it dry and bright, and replacing it at once. There is then no fear of forgetting it.

It is important in choosing fern shades of this description to see that the glass dome fits *loosely* in the pan which accompanies it. One of our shades, which was a tight fit, was one day removed into a sunny window for a few hours to make room for some domestic operations. The sun heated the air within the shade, the expanded air had no means to escape, and it burst the shade with a loud explosion into a multitude of fragments. A guinea's worth of glass was thus lost in a moment, and a collection of Selaginellas placed in jeopardy through neglect of this precaution.

Success in these matters often turns on points of management that appears trifling; let me, therefore, describe the process of planting a fern shade. If intended for a winter ornament, it should be planted in July or August, that the ferns may be established before the decline of the season, and if they are evergreen kinds they will have plenty of time to throw up an abundance of fine fronds, which the liberal supply of water from below, with regular ventilation, will render luxurious and beautiful; and before winter comes, the excess of moisture will be gone, but the soil will hold enough to render watering almost unnecessary until spring. In a large pan—say, six inches depth—lay down two and a half inches of broken flower-pots or cinders of the size of walnuts; on this lay a thin coating of half-decayed moss or sphagnum. Fresh green moss is apt to go sour or breed fungi, and therefore it is preferable if it has been for some time exposed to the action of moisture. Fill up to the level of the rim with a mixture of turfy peat, small broken charcoal, and the siftings of broken pots, varying from the size of a hazel-nut to that of a pea, with plenty of silver-sand. I never measure the ingredients of any compost, but the beginner may like to be saved from doubt, and therefore let the proportions be taken as follows:—Peat three parts, silver-sand one part, broken charcoal and crock-siftings one part. This compost should be broken up and mixed with the hand, and should be in a free lumpy state. Ferns will never prosper if the compost is sifted, but a little of the finest of it should be put aside to dress the surface with when the planting is completed. Now, take a can of boiling water, and water the soil till you have supplied enough to rise to the top of the drainage. The water should be poured into the centre first to warm the soil gradually; poured against the glass suddenly it may shatter it. I have used the boiling water now for many years on every occasion of planting a fern case, and have not yet had one accident. With a little caution there is no risk. The use of the boiling water is to destroy every insect that may have escaped your eye when breaking up the peat. It will not only do that, but kill their eggs also, and equally make an end of the seeds of weeds and the mycelium of fungi; all of which are enemies better got rid of at first than to be hunted for when their ravages become a source of alarm. The over cautious may of course scald or bake the materials before filling the pan; in that case they must not be put in the pan until nearly dry again.

When the pan is nearly cold the ferns may be planted, and the process of planting will consolidate the compost, so that it will, when all is finished, be an inch below the edge of the pan, as it ought to be; it may indeed go below that, and need filling up with some of the finest of the mixture, which should be sprinkled over as a finishing touch.

In any case of difficulty in obtaining peat of a friable and fibrous texture for fern cases, a mixture of equal parts of cocoa-nut fibre refuse and charcoal dust will answer admirably.

I could enumerate fifty groups of ferns offhand suitable for bell-glasses, but tastes differ, and the best possible way to please yourself is first to obtain a sufficient number of ferns of *suitable kinds* and arrange them as you think best. I will, however, as my journal of fern work is at hand while writing this, give you the planting of a bell-glass measuring twenty-two inches across which I once planted for a friend who knew well how to manage these things, and who was pleased to say that, though very fastidious on matters of taste, she was well satisfied with my way of doing things.

In the centre *Cheilanthes farinosa*, the most accommodating of all the silver ferns. At regular distances round it *Adiantopsis radiata*, *Cheilanthes tomentosa*, *Asplenium fragrans*, *Asplenium vivipara*, *Pteris argentea*, a little silvery gem, *Elaphoglossum brevipes*, *Doodia lunulata*. All over the surface, so as to quite cover it, *Selaginella apoda*.

The following are six beautiful ferns adapted for glass shades in the hands of beginners; in fact, if they are not drowned with water, and have but a moderate amount of light, they are sure to thrive even if neglected for weeks together. *Asplenium marinum,* the sea spleenwort; *Doodia caudata, Scolopendrium vulgare ramo-marginatum,* a tasselled variety of hartstongue; *Asplenium viride,* the green spleenwort; *Adiantum setulosum, Lomaria lanceolata.*

Let us now consider the Fern case proper, and first as to how it should be made.

Elegance is a prime requisite, but as tastes differ we shall say but little on that point. The lighter the structure consistent with safety the better. We do meet with very ugly fern cases at times, and the ferns within them are usually in a bad state of health. The fact is, heavy framework and cumbrous ornaments *obstruct the light,* and therefore ugly fern cases are, as a rule, to be condemned for that reason, if for no other.

A simple rectangular figure as indicated in the simple sketch annexed is undoubtedly the best for all general purposes; moreover, this box-like form may be made the basis of an elaborate design: out of it may rise a miniature mosque or a Crystal Palace, as I have shown by figures in the chapter on Fern cases in "Rustic Adornments." A figure of the best fern case we have ever had was published in the 1st volume of the "Floral World;" it is a handsome case made of Ransome's imperishable stone, surmounted by a tall lantern.

At every step in designing and constructing it must be borne in mind that ferns are to be grown in the case, and, therefore, it must afford access to light and air, and egress for water.

Accessibility to the ferns is of the utmost importance. If the case is small it should be possible to lift off the whole of the glass framework at any time. If too large

RECTANGULAR FERN CASE.

for that there should be doors on two sides, because in reaching across from one side to plant a fern on the opposite side some mischief may be done. The case figured on p. 50 may be taken to pieces in a few seconds, as each sheet of glass is fitted in a separate frame, and all the frames drop into grooves and are braced together at the corners by means of small hooks and eyes. The patent cases made by Gray of Danvers Street, Chelsea, are of this make, and they have the additional advantage of a boiler to afford warmth from below; this boiler requires to be filled only once or twice a day in winter according to the severity of the weather. Ventilation is easily effected without causing a draught by simply tilting up the top glass.

It is a great convenience if a fern case can be moved about without difficulty, and it is astonishing what may be accomplished in this respect by the exercise of forethought. For example, if you buy one of Gray's cases it will be supplied on a miserable set of legs with wooden castors, and even if a small one it will be difficult to move it. But if you follow our plan the difficulty vanishes, and you may take your fern cases with you on your travels, or at all events wheel them from room to room with a mere touch. The legs sent with the case are converted into firewood, and the case is put on a strong framework made by our own carpenter, of which the annexed figure affords an accurate representation, save and excepting one particular.

The frame figured is one on which stands a case measuring three feet long, two feet high, and eighteen inches wide. The frame consists of a skirting-board, A, with neatly-moulded top edge, six inches in depth, mounted on four neat but strong legs, which are fitted with large brass castors, all wooden and iron castors being rubbish. From the ground to the top edge of the skirting-board the measurement is seventeen inches. The case does not stand *on* this frame, but *in* it, that is to say, it rests on the half-inch ledge, B, which extends all round inside, and which is added to at the corners by the blocks, C, which are placed there to increase the strength of the frame. The advantage of this mode of mounting is not in appearance only, though that is of some importance in an article intended for the adornment of a chamber. One important advantage is the ease with which the case can be moved about; an immoveable case is a nuisance except in some peculiar circumstanees. The engraver has forgotten to add the castors.

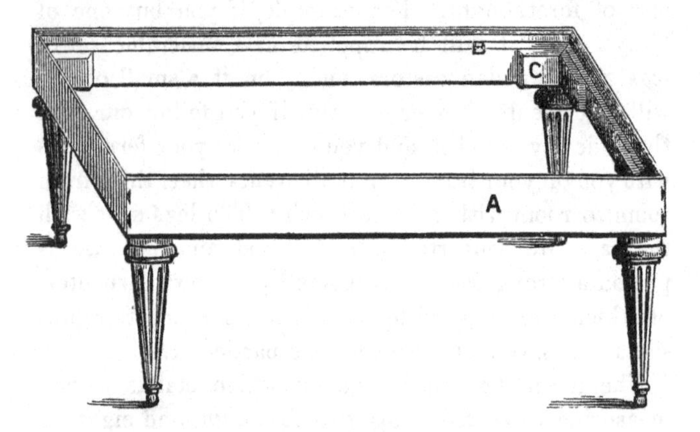

That there should be means of escape for surplus water is desirable, but not absolutely necessary. The experienced cultivator will never allow a fern case to become so saturated with water as to be hurtful to the health of the ferns; but the beginner is almost sure to fall into this error, and the first disaster that occurs is, nine times in every ten, to be attributed to a water-logged condition of the roots. Make it a point to ascertain, when purchasing a fern case, if there is any perforation of the base to allow of the escape of water. If there is none you must be the more cautious to avoid charging the soil with excess of moisture. It is thought to be impossible to provide drainage in those cases which have boilers beneath, but I think it may be done, and I hope some day to find time to carry out my views.

For the benefit of mechanical and experimental readers, here is my idea of a fern case combining means of heating with effectual drainage.

I would go to the expense of having all the metal work in copper well tinned; it would be expensive, but would last for ever. A, should be a trough for soil, resting on a ledge all round the inside of the outer wooden casing, and admitting of being lifted out at any time. For the lifting there should be a ring attached on each of the four sides. In the centre of this I would insert a pipe, F, for escape of drainage, and this very simple process makes an end of the principal difficulty. The bottom of the trough might slope down every way to the pipe, F, which would render its action more effectual. For the communication of heat I would have a space, B, allowing a depth of two inches at least under the bottom of the trough, and additional spaces under the sloping ends of the trough. By increasing the quantity of water so as to fill the ends as well as the bottom, a maximum of heat would be obtained. Now, to fill this reservoir need not be so ridiculous an affair as it is at present; my idea of the matter is to have a whole side of the wooden frame removable at a touch, so that we could get to the reservoir and fill it with as much ease as one might fill a washing-tub. I have shown a removable portion only of the end C. I must leave it to the imagination of the inventive reader to work out this point, confident that he will have no

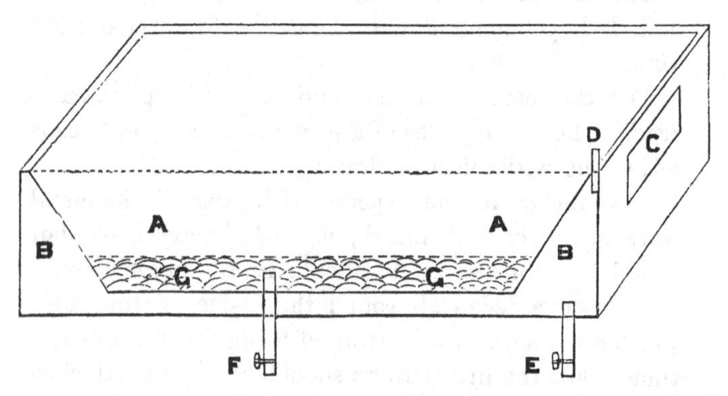

A, trough containing soil for ferns; B, reservoir for hot water; C, opening for filling reservoir; D, air pipe; E, tap to draw off water from reservoir; F, tap to draw off drainage water from soil; G, stratum of crocks for drainage.

difficulty in opening the side of the case so as to pour water into the reservoir with some speed from a large can, instead of dribbling it in as now in a way that suggests that fern-growers ought to live for ever if only for the sake of keeping their cases warm. The pipe, F, would have to be a fixture, with a close-fitting india-rubber ring surrounding it where it enters amongst the crocks at the base of the trough, A; this, of course, prevents the water from B rising up amongst the soil and flooding the ferns.

ROSHER'S FERN PILLAR.

CHAPTER IX.

MANAGEMENT OF FERN CASES.

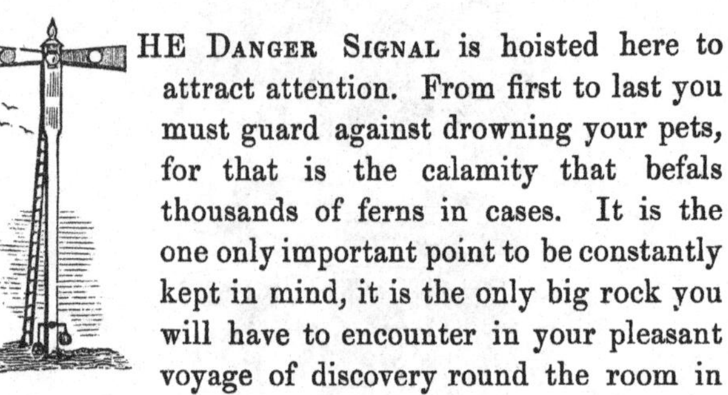

HE Danger Signal is hoisted here to attract attention. From first to last you must guard against drowning your pets, for that is the calamity that befals thousands of ferns in cases. It is the one only important point to be constantly kept in mind, it is the only big rock you will have to encounter in your pleasant voyage of discovery round the room in search of fern island. So long as the soil is moderately damp water need not be given to the roots; but at almost any time a slight shower over the fronds by means of a syringe will be beneficial. In winter the syringe must be cautiously used, and if there are any gold or silver ferns in the case, care should be taken to prevent a single drop of water falling upon them, as the farina with which they are covered ought never to be washed off. Air must be given regularly and with judgment; a brisk breeze will do much mischief, and as dust is to be kept out as much as possible, do not open your case while sweeping is going on, or when, through open doors and windows, a young hurricane is enjoying his gambols.

Beautiful effects may be produced by a judicious use

of mimic rockeries, and they are useful as affording elevated and well-drained sites for small ferns of delicate growth. The best material for constructing rockeries, arches, &c., is common coke. It adds but little to the weight, and it may be made to look like stone by soaking it with water and sprinkling it with Roman or Portland cement. For the formation of irregular mounds and to dot about amongst the ferns to vary the surface, soft sandstone or rough and rather soft pieces of brick burrs should be preferred, not only on account of their suitable colours, but because they soon get coated with natural growths of moss and add much to the beauty of the little garden. But the grand thing is to have a sufficiency of healthy ferns of handsome varieties, everything else must be made subsidiary to that desideratum. Have good ferns and grow them well, and you will not be greatly exercised about the niceties of gimcrackery.

Vermin of many kinds occur in fern cases in spite of all precautions; mysterious nibblings of fronds are noticed, sometimes the crown of a valuable plant will be found eaten away. The marauders may be woodlice, slugs, or the larvæ of small beetles. Trap them, if possible, by inserting fresh lettuce leaves in the chinks you suspect they frequent. Or place slices of fresh apple under tufts of moss. Examine the baits daily, and keep them always fresh. If you can put a few glowworms in a case infested with vermin, there will be a rapid clearance made; toads are good vermin killers, but they do not add to the beauties of the scene, and they are apt to squat on the tender rising fronds of some delicate fern, and do more harm than good. Green fly, or aphis, is rarely seen in fern cases, and when it occurs it is usually a sign that there has been neglect in giving air. The best way to remove

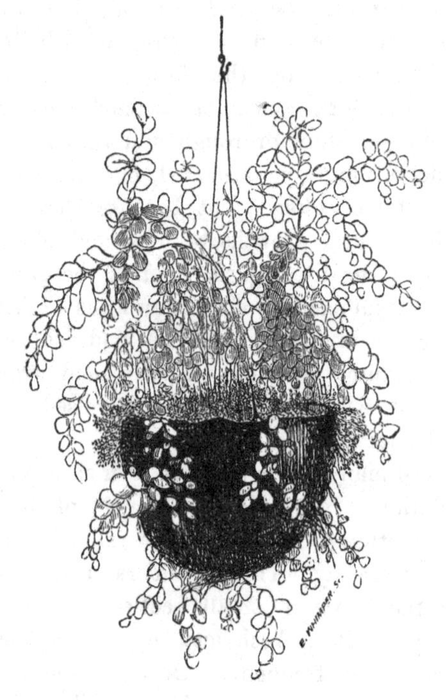

ADIANTUM SETULOSUM.

the aphis is by means of a soft brush or camel's-hair pencil, and to prevent its recurrence give more air.

For suspending ferns in cases, the outside husk of the cocoa-nut may be used, and also the hard inner shell. For general purposes the latter is preferable.

If broken with a clean edge half a shell makes a capital basket. It requires a sharp saw and some patience to cut the edge nicely if it is much jagged. The holes must be burnt in the shell, as they are apt to split if any attempt

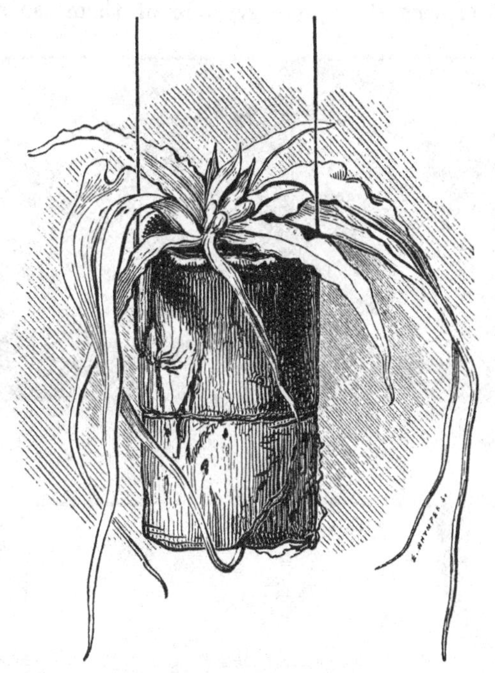

CAMPTOSORUS RHIZOPHYLLUS.

is made to pierce them with awl or gimlet, and, besides that, they are so hard that the best bit of steel breaks like glass against them. Make three holes for drainage, and two very small holes near the rim, exactly opposite each other, and use for suspending a silver string of the Spanish guitar, which will never rot, and is as soft and pliable as packthread. The two ferns here represented are examples of my mode of suspending. The Adiantum is in a cocoa-nut, bored all over with holes a quarter or third of an inch in diameter, and it has pushed crowns through every one of them, so as to

smother the outside with foliage. The Camptosorus is in a little case formed of thin bark, bound with brass wire. To water these, the best way is to lift them out and lower them into a deep vessel, with a stick passed through the suspending cord and laid across the top of

ACROSTICHUM QUERCIFOLIUM.

the vessel, so that they can sink the full length of the cord and be thoroughly saturated. They can be lifted out in a quarter of an hour, and allowed to drip for a few minutes by again lodging the stick at each end in a suitable place.

THIRTY FINE FERNS FOR CASES WITH ARTIFICIAL HEAT. The best for beginners marked thus *.

Anemidictyon phyllitidis,* 9 inches; Aneimia adiantifolia,* 9 in.; Asplenium bifidum, 18 in.; A. fragrans, 9 in.; A. heterodon, 12 in.; A. radicans, 9 in.; A. Mexicanum,* 6 in.; A. polymorphum,* 4 in.; Blechnum lanceolum,* 4 in.; B. intermedium, 6 in.; Campyloneurum lucidum, 12 in.; Cheilanthes micromera, 12 in.; Diplazium radicans, 9 in.; Elaphoglossum brevipes, 6 in.; Fadyenia prolifera, 3 in.; Acrostichum quercifolium,* 4 in.; Polypodium loriceum, 12 in.; Hemionitis cordifolia,* 4 in.; Hymenolepis spicata,* 9 in.; Hypolepis tenuifolia deformis, 12 in.; Litobrochia leptophylla, 18 in.; L. pedata, 6 in.; Lomaria attenuata, 12 in.; L. Patersoni, 9 in.; Nothochlæna vestita, 6 in.; N. tenera, 6 in.; Olfersia cervina,* 18 in.; Pleopeltis percussa,* 12 in.; P. membranacea,* 12 in.; Pteris calomelanos,* 6 in.

THIRTY FINE FERNS FOR CASES WITHOUT ARTIFICIAL HEAT. The best for beginners marked thus *.

Asplenium appendiculatum,* 12 in.; A. attenuatum,* 4; A. crenulatum, 18; A. dimidiatum, 9 in.; A. nitidum, 6 in.; Adiantum tinctum, 12 in.; A. assimile,* 9 in.; A. cuneatum,* 18 in.; A. cristatum, 9 in.; A. formosum,* 18 in.; A. fulvum, 12 in.; Doodia aspera,* 12 in.; D. caudata,* 6 in.; D. lunulata,* 9 in.; Lastrea acuminata,* 8 in.; L. glabella, 8 in.; Niphobolus lingua, 9 in.; N. pertusus, 6 in.; Nephrolepis pectinata, 18 in.; N. exaltata,* 30 in.; Onychium Japonicum,* 15 in.; Platyloma rotundifolia, 18 in.; Pleopeltis pustulata, 9 in.; Phlebodium aureum,* 36 in.; P. sporodocarpum,* 30 in.; Polystichum triangularum, 6 in.; Pteris crenata, 12 in.; P. geraniifolia, 9 in.; Pteris cretica albo-lineata,* 18 in.; P. heterophylla, 6 in.

TWENTY-FOUR FINE FERNS FOR SUSPENDING IN CASES. The best for beginners marked thus *.

Adiantum setulosum,* 6 inches; Asplenium brachypteron,* 6 in.; A. flabellifolium,* 9 in.; A. pinnatifidum, 6 in.; A. reclinatum, 6 in.; Camptosorus rhizophyllus,* 5 in.; Cheilanthes sieberi, 10 in.; Davallia pentaphylla, 9 in.; D. bullata, 6 in.; D. dissecta, 18 in.; D. elegans, 12 in.; D. solida, 6 in.; D. pyxidata, 18 in.; D. canariensis, 12 in.; D. decora, 6 in.; Humata heterophylla, 6 in.; Hypolepis amaurorachis, 18 in.; Niphobolus lingua, 9 in.; Nothochlæna nivea, 6 in.; Oleandra nodosa, 6 in.; Pleopeltis lycopodioides, 3 in.; P. stigmatica, 6 in.; Polypodium rugulosum,* 9 in.; Pteris scaberula,* 12 in.

FORTY SMALL-GROWING CASE FERNS, SUITABLE FOR A FERN PILLAR OR ROCKERY.

British—Adiantum capillus veneris, Asplenium fontanum,* Asplenium germanicum,* Cystopteris regia* (deciduous), Polypodium dryopteris (deciduous), Scolopendrium vulgare bimarginata cordatum,* S. v. cristatum minus, S. v. divergens, S. v. geminum, S. v. glomerato-digitatum, S. v. lacerato-marginatum, S. v. proliferum, S. v. ramo-marginatum,* S. v. ramo-

CYSTOPTERIS REGIA.

proliferum, S. v. ramosum, S. v. Wardii,* Woodsia alpina (deciduous), Woodsia ilvensis * (deciduous).—

Exotic—Camptosorus rhizophyllus,* Lomaria alpina,* Acrophorus hispidus, Adiantum setulosum,* Asplenium flabellifolium,* Asplenium attenuatum,* Asplenium brachypteron, Asplenium nitidum,* Asplenium obtusatum, Asplenium pumilum, Asplenium pinnatifidum, Blechnum intermedium,* Campyloneurum cæspitosum, Diplazium plantagineum, Davallia decora, Elaphoglossum brevipes,* Gymnopteris quercifolia, Goniopteris scolopendrioides, Lomaria lanceolata,* Nothochlæna vestita, Nothochlæna tenera,* Pleopeltis stigmatica, Doodia caudata.*

PTERIS TERNIFOLIUM.

CHAPTER X.

THE ART OF MULTIPLYING FERNS.

THERE are two modes of increasing ferns—by division and by spores. Both plans are easy enough up to a certain point, but we need not trouble ourselves about the point at which serious difficulty commences, for in truth no beginner should be troubled on that score. I will suppose you have a large plant of the common Male fern (*Lastrea filix mas*) or of the common Hartstongue (*Scolopendrium vulgare*), and you wish to make more of it at once. The best time to operate is when the fronds are just rising in the spring, but it may be done at any time if proper care be taken. We take the plant out of its pot, or lift it out of the ground by means of a fork or trowel, and lay it on a board or table. Probably at a glance you will discover that a number of distinct crowns, each with a tuft of roots attached, may be easily removed from the outside by the use of a strong sharp knife. Separate such offsets, carefully disentangle their roots from the mass, and at once pot them in very small pots in the sort of mixture already advised for use in growing pot ferns in Chapter VI. Prepare the pots by putting in them plenty of small crocks for drainage, over them a thin wisp of dry moss, or a bit of fibre torn from the peat, then put the little plant in its place and fill in round the roots and press moderately firm. If this is done in spring before the fronds have unrolled, you may be content to give a little water and put them in a frame and keep shut rather close until they begin to grow, giving very little water until they have made some progress. If you have no frame, the pots may be placed in any sheltered shady corner on a bed of coal ashes, and will almost take care of themselves. But the old plant remains, not much diminished in size by the removal of the offsets. Lay it on its side and carefully pass the knife through the centre of it, and as it separates into two portions, you will probably see how to divide it yet further without using the knife again, securing to each piece a *centre* or crown, and a tuft of roots. Treat these in the same manner as the offsets; or, if you have not rashly torn the plant to shreds, you may at once plant the divisions in the fernery, filling in round its roots with your best mixture of sandy peat, and pressing each firmly in its place. A little shade, and occasional sprinkling, will assist them to become established, and they will soon take care of themselves.

If you were to proceed in a similar manner with the same plants in the summer time, when crowned with luxurious leafage, you would have to be as quick as possible about the work, and pot all the pieces and shut them up in a frame for a fortnight, to recover and make fresh roots, during that time taking care to sprinkle them frequently and also to avoid making the soil in the pots very wet, for too much moisture to roots of any kind that have been disturbed and need time to regain their wonted action is like poison.

So much for the division of ferns that form clustering crowns. Let us now take a tuft of common polypody. Here we find a quantity of fleshy rhizomes of the thickness of a lead pencil entangled amongst masses of fibrous roots. You may cut or pull to pieces this tuft almost *ad lib.*, provided each separate position has its own roots reserved to it. The pieces must be potted rather differently to the others, as their roots run upon the surface chiefly, and they thrive best in a moist spongy material. The surest way to make plants of them will be to prepare the pots by putting in at least one third depth of crocks, then nearly fill them with sandy peat, and on that spread a little cocoa-nut fibre to make a soft bed; then lay one of the pieces on the bed, put some more cocoa-nut fibre over it almost to bury it, and press it down firmly. Water and place in frame and treat as in the first practice. They will soon begin to grow, and will want no particular care after a few weeks.

Now, by these two methods may the greater part of all known ferns be multiplied; there are exceptions, as in the case of tree ferns, for example, but the exceptions are few. Those that grow in clustering crowns may be divided as in the first practice, those that extend by creeping rhizomes may be cut to pieces as in the second practice.

As you extend your operations, you will not be long in discovering how easy it is to kill ferns by one or the other of these processes. For general guidance I will say, then, be sure before you begin that you know what you ought to do. If you cannot see how to divide a plant without spoiling it because it neither offers you offsets, nor a crown large enough to be severed without danger, leave it alone, be content and wait. The habits of different species must be observed also if the cultivator would become expert in propagating. Take for example *Onoclea sensibilis,* a charming flowering fern for a damp place in a rockery out of doors or under glass, which I hope you will obtain at the first opportunity, if you do not already possess it. Now, this fern propagates itself; that is to say, the rhizoma runs along near the surface, and at some distance from the parent plant throws up several distinct crowns. Leave the plant alone for a couple of seasons and it will be surrounded by, or rather it will consist of, a number of separate centres of growth forming a large rich mass of vegetation. You may divide this into as many pieces as you please, provided each piece has its own centre and tuft of roots, and make plants of them all with patience aided by shade and moisture. Take on the other hand a potted Gleichenia that has been in the same pot two or three years, and you will find it dead in the centre, but all round next the pot will be a series of crowns. Carefully knock it out of the pot, lay it on its side, pass the knife through it, separate the pieces and shake from them the old worn-out soil and pot as before; but in this case put the plant into a warm pit or some other place where it can have a temperature of 60 to 70° with shade and a humid atmosphere, to encourage a new growth. The principle is the same in every case, but as different classes of ferns differ in constitution, so the practice must be varied to suit them.

In every case of multiplying by division it must be borne in mind that the operation severely taxes the energies of the plants, hence the need of extra care for some time afterwards to restore their vigour. The soil in which small offsets are potted may with advantage contain more sand than strong plants require, and it may be quite fine in texture, whereas for strong plants it is best somewhat lumpy. So, again, extra warmth and occasional damping of the crowns, and a humid atmosphere with shade from sunshine, are aids of great importance. Begin with cheap hardy kinds, and take as much pains with them as you would with the most tender and costly, and you will enjoy the work, be rewarded with success, and acquire experience for higher flights in a most amusing pastime.

> "If at first you don't succeed,
> Try, try, try again."

Now for the spores; and first by way of preface. The spores of tropical or hothouse ferns must be placed in heat or they will not germinate. The spores of greenhouse ferns may be raised in summer time without the aid of artificial heat, but it is a safer plan to put them into a propagating house and treat them the same as the tropical kinds until the little plants resulting from their germination have made some progress. As for the spores of hardy ferns, they may be raised in a frame kept close and shaded.

First secure some large shallow pans, and bell-glasses to fit them. Of course common flower-pots will answer the purpose, but large shallow pans are better. Nearly fill the pans (or pots) with broken flower-pots, the top stratum of which should be broken to the size of peas. Sweep all the dust made in breaking the pots into the pans with the smallest of the crocks, and then put in an inch depth of a mixture of equal parts fine peat and silver sand. Water with a fine rose, and if the watering washes the fine stuff down, and causes the points of the small broken pots to peep through, all the better—that is as it should be. Now take a *ripe* frond of a fern on which there is plenty of fruit, and while holding over the prepared pan, sweep the hand over it, or tap it smartly, and you will see the fine dust—the veritable fern seeds—fall freely. Regulate your movements so as to scatter the dust all over the surface, and then put on the bell-glass.

The proper place for pans so prepared is wherever they can be kept warm and dark, and yet be within sight, so that they are not neglected. They must be kept always moderately moist, but never wet, and as watering with a water-pot would simply wash the seed away, follow the neater practice of placing the pans in vessels of water. If they are immersed in one inch depth for an hour, the whole mass will become moist throughout by capillary attraction, and not a grain of sand or seed need be moved from its place.

Have patience, and you will see first a film of green confervæ, which is a good sign, next little leafy growths, resembling the liver worts or marchantias. By-and-bye from these curious leafy things little fern fronds will rise, and you will know thereby that you did not sow the seed in vain.

Be in no hurry to disturb the little plants. More and more will appear; they will crowd and jostle one another, and they will form a sort of microscopic forest, and very likely will appear to be very different in form to the frond from which the seed was taken, for they do not usually acquire their true characters until they have made some advance. The time will come at last to give them more room, but before you disturb them remove the bell-glass, and habituate them to the enjoyment of more air and light than they had in their earliest infancy. I usually allow seedlings to remain a whole year in the seed pans, and then pot them off, and this plan will be found a safe and good one for general adoption.

The process of potting consists in lifting each little plant with its tuft of roots unhurt into a very small pot nearly filled with a mixture of fine peat and sand, and then covering its roots with the same material, and tucking it in comfortably. Shut them up in a frame in a greenhouse, or put them close together under large bell-glasses; by some means or other keep them comparatively warm and shaded; give gentle sprinklings or rather *dewings* over their leaves, and but little water to the roots, and they will soon grow and become bonny little plants.

In a rather dark and damp corner of one of my greenhouses I have a glass frame on a stand which is used expressly as a nursery for seedling ferns. You might make one for yourself by taking a shallow box, and covering it with sheets of stout glass. Make a bed inside the box of a few inches depth of cocoa-nut fibre, or silver sand, or clean small pebbles, and on this bed place the little pots and put the glass over. You have complete command over them by this plan to kill them by excess of shade and moisture, or by exposing them to sunshine; or to make them grow by giving moisture and shade enough to keep them in the first instance, and to admit more light and air, to strengthen them as they advance and become strong enough to be shifted into larger pots. Small fern cases with moveable tops make admirable nurseries for seedlings when they are grown in sitting-rooms.

We have spoken of ferns that run about and multiply by means of their rhizomas. A parallel case is seen in ferns that shed their spores, and sprout up into life without aid from any one, and almost anywhere. It will be amongst your earliest surprises and delights in fern growing to find seedlings in your fern cases, on the banks, and walls, and stones, and even pavements of your fern-houses, and in crevices of the rockery out of doors. Some ferns increase spontaneously with such freedom as to become weeds, but the wise man will not despise them on that account. He will be quickened in love and thankfulness to God for making beauty so cheap on the face of the earth. He will rejoice that the humblest and least enlightened cannot fail to see that in the mystery of life is afforded us deep and blessed impressions of the direct relationship of the Divine nature to the manifestations of the Divine will in visible things.

To the observant mind there is nothing trivial or paltry in nature, and the growth of a fern seed is the beginning of a mysterious life, the end of which no man can predicate or understand.

> Behold! we know not anything;
> I can but trust that good shall fall
> At last—far off—at last, to all,
> And every winter change to spring.

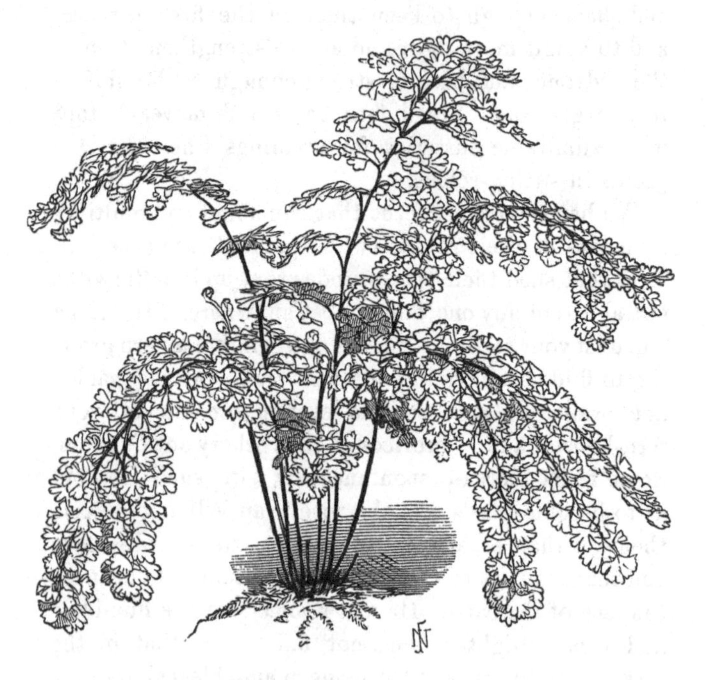

ADIANTUM EXCISUM MULTIFIDUM.

CHAPTER XI.

BRITISH FERNS.

THE number of known ferns is about 3000. How many are unknown we cannot even rudely guess. The British species number 46; many of these present us with varieties in great abundance, that is to say, with forms differing from their types (or what we regard as types), and these varieties number full 500, and no living person possesses the whole of them. It is not the business of this book to treat of ferns botanically, nor to speak of the British ferns exclusively, yet it would hardly be complete—restricted as its object is—unless it contained at least one chapter on the Filices of Britain, more especially as many persons only cultivate the British ferns, and find enough to amuse them in the fern way in making collections of native species and varieties. Let no one suppose a *complete collection* to be desired, for it is not, except for strictly scientific purposes. It matters little for our purpose whether it be desirable or not, for the fact is, a complete collection has never been formed and never can be. My excellent friend Mr. Sim, of Foots Cray, Kent, enumerates in his last catalogue 365 British species and varieties in all—enough for us to choose from for the materials for a fern garden. If the reader has no innate horror of statistics, a few figures may be interesting. It must be understood that amongst the *varieties* are many extremely curious plants. Some are richly tasselled and fringed, some have duplicated fronds, and the variations otherwise comprise imitations (or resemblances to) stag's horns, frills, fans, wires, bristles, embroidery, braiding, puckering, and embossing. Some of the varieties are notched as if a child had cut faces out of them, others are shrunk up to mere stalks; some have spores on the wrong side, that is to say, on upper side of the fronds, others never produce spores at all, and a few produce their offspring ready made in the form of little plants at the points of their fronds or on every part of their leafy surface. Some varieties are so curious, so rare, and so difficult to multiply that they range in price from one to five guineas a plant. This need not terrify the humble fern collector, for many of the handsomest may be bought for a shilling each. The catalogue prices of 319 kinds enumerated in Sim's catalogue amount to £130 16*s.*—say if those not priced be added, £200 for one plant each of the 365. The varieties of hartstongue alone are about 100 in number, and to buy one each would cost in the aggregate £50. Here ends the statistical statement. Now let us hastily run through the list of British ferns, saying nothing about synonyms or knotty points in classification, for with these matters we cannot now have anything to do. For our purpose an alphabetical arrangement will be best.

Adiantum.—*A. capillus veneris*, the true maiden-

ADIANTUM CAPILLUS-VENERIS.

hair. There are a few varieties, but we need not enumerate them. The requisites for the growth of this lovely fern are warmth, shade, and moisture. In the damp and rather dark parts of a plant stove it soon becomes a weed, and sows itself by spores on bricks, stones, wood—anywhere. I have had it grow to perfection between the bricks inside a well. To have a plant in a room, the best way is to appropriate to its use a fifteen-inch bell-glass, fitted to an earthen pan of red flower-pot ware. The soil should be sandy peat, with a fourth part of broken flower-pots or soft broken stone added. Give air daily for half an hour; never leave the glass off and forget it; do not saturate it with moisture, and—*have patience.*

ALLOSORUS.—*A. crispus,* the mountain parsley fern. Coddling will kill it. It loves fresh air; will grow amongst pebbles or broken stone with a little sandy peat to give it a start. Shade is good for it, but I have seen it growing gloriously in the full sun. *Beware* if there is one snail in the garden; catch him and throw him over the wall into the next garden, or he will gobble up your plant as a cat would a mouse. It is a good plan to put a bell-glass over a newly planted piece to protect it from the vermin; the glass can be taken away when the plant has grown a bit.

ASPLENIUM.—*A. marinum* is one of the best case ferns known. It loves sand and stone, and warmth and vapour. To plant it in the open air rockery is a risk, but it will do well in the cool fern-house near the floor.

A. trichomanes and *A. viride* are charming ferns to

plant in a cool house or a case, or in sheltered chinks in the open rockery. If it should ever speak it would be in such words as once startled the horticultural community, "Give me air or I shall die." Soil to be bricky and sandy; fat peat is poison to it.

A. fontanum—a gem for the case.

A. ruta-muraria.—Stagnant moisture is ruin to it; plant with the crown quite above the surface; soil one half broken brick or stone, the other half very sandy peat. A lovely fern for planting in a chink in an old wall in a shady sheltered spot.

A. septentrionale, a difficult fern to grow. Try it in a pot in a frame, in soil three parts sand and soft stone, and guard it with fear and suspicion against slugs, snails, and woodlice.

ATHYRIUM.—*A. filix-fœmina* is the Lady fern, and well deserves the title. Please excuse description or eulogy; see it and believe. It will grow anywhere under glass, or in the open air, if in a shady moist position. I have a grand plant growing in the gravel walk at the foot of the bastion, and more than I can count in other places. A fine pot fern, growing well in fat peat or in common loam, with sand, or in any soil not chalky, with the help of a little cocoa-nut fibre to mellow it. Be sure to drain the pot effectually and give plenty of water. Oh, how it will smile upon you if you treat it kindly!

The following varieties are fine—*Coronans, Corymbiferum, diffuso-multifidum, Elworthi, Fieldiæ, Frizelliæ, grandiceps, Grantiæ, multifidum.*

BLECHNUM.—*B. spicans*, the hard fern, is a noble

and very distinct fern. Try your hand at a large pot specimen—when four or five years old it will be grand. A rather strong soil, with good drainage, suits it; say yellow loam three parts, leaf mould two parts, and grit

ATHYRIUM FILIX-FEMINA, *var.* CORYMBIFERUM.

obtained by sifting the sweepings of the gravel walks one part. By the way, this is a capital plan of obtaining clean sharp sand. We rarely buy sand, as we sift all our sweepings and spend the sand money in keeping the gravel perfect.

The following varieties are good—*imbricatum, lancifolium, multifurcatum, ramosum, strictum.*

Ceterach.—*C. officinarum,* the scaly spleenwort, is

ATHYRIUM FILIX-FŒMINA GRANDICEPS.

a very interesting fern. It grows luxuriantly in our cool fern-house, in a chink of the rough wall near the door. A good pot fern. It loves air, stone, old mortar,

GYMNOGRAMMA LEPTOPHYLLA.

shade, and *perfect drainage ;* try it as an aquatic, and say "farewell" to it before you begin.

CYSTOPTERIS.—*C. fragilis,* the brittle bladder fern, is a little dear: treat it kindly; it will do in a snug nook of the open rockery, but it must be in front because small. The varieties are not worth having. *C. montana,* the mountain bladder fern, and *C. regia,* are gems. These do best in the open air or cool greenhouse. They need shade and shelter, but love fresh air. Prepare a bed a foot square by removing the soil a foot deep. Then partially fill up with broken bricks and charcoal, and upon this bed place four inches depth of a mixture consisting of equal parts peat, silver sand, the finest dust of cocoa-nut fibre refuse, and soft silky loam. Place the plant in the centre of the bed, close the soil firmly around it, and put a bell-glass over. Take off the bell-glass every morning, and wipe it quite dry, and place it over the plant again. Keep the soil moist, and in due time the plant will grow. After six months of such nursing, it will take care of itself in every respect except one, and that is, it will invite the attacks of snails and slugs, which are very fond of it. These must be trapped and destroyed with energy; you must be a Thug to such people.

GYMNOGRAMMA.—*G. leptophylla* is the only Britisher of this lovely family. This little gem is an annual. To secure it for ever, get a plant in a pot, and keep it in a fern-house or shady moist pit. It will shed its spores, and the parent plant will perish. The next season it will appear plentifully as a weed on bricks, stones, borders, &c., &c. Pot a few to give away, and allow the remainder to attain maturity and shed their spores for the next season.

HYMENOPHYLLUM.—*H. Tunbridgense* is the Tunbridge filmy fern, a cynosure, a paragon, a paradox. It represents a race, all of which require similar treatment. They are all easily grown if dealt with in a proper manner in the first instance. Suppose we consider how to grow a nice patch of any of them. Get a large earthenware pan (flower-pot ware) and a bell-glass to fit fairly within the rim. A fifteen-inch glass would be best, but one half that size will do to begin with. Spread over the bottom of the pan a layer of broken pots, then lay down a bed of very sandy peat—say peat and silver sand equal parts. On this bed place some blocks of stone of the size of the fist, and less, and press them down, and fill in between them with the same mixture of peat and sand. Make all this quite firm—make it, in fact, *hard.* Now draw out a small stone, and introduce the plant, spreading out its black hair-like roots, which cover with the mixture, and bed it in close, so that it will sit, so to speak, close to the general surface of the stone. If you can plant little pieces all over the pan between the stones, you may get the pan filled more quickly, but it is a risk for a beginner to tear up a plant as a practised hand would do. Wet the whole by means of a fine syringe; place the bell-glass on, and press it slightly so as to make it fit pretty close, and place the pan in a warm room near the window, or in a snug, warm, shady corner of the greenhouse, or in a cool part of the stove, and do not look at it for a week; then take off the glass and give another gentle sprinkle,

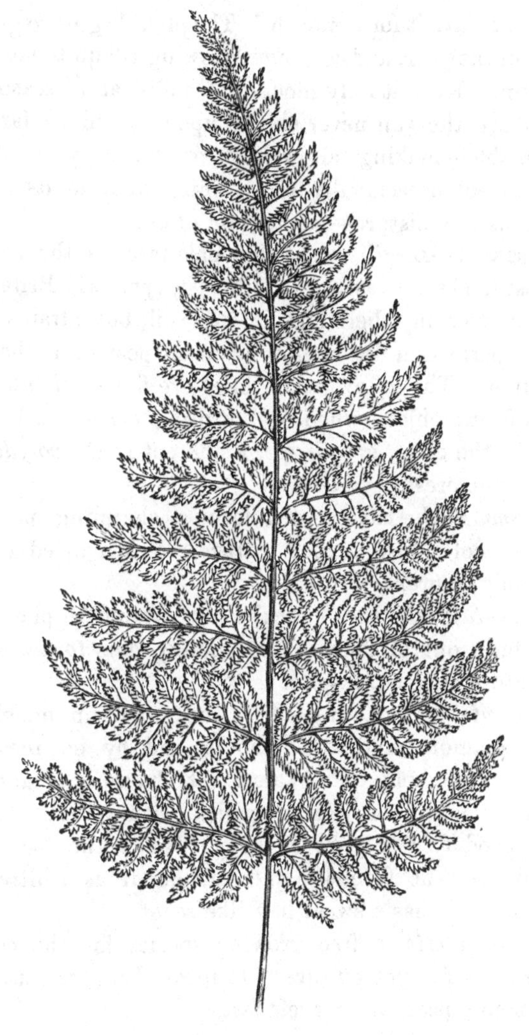

LASTREA ÆMULA.

and next leave it for a month. The plant begins to grow early in the year, and continues growing till quite late in autumn. Keep it only moderately moist at all seasons. Take care the sun never shines upon it, and as far as is possible—making allowances for curiosity, or the necessity of occasionally ascertaining what is its condition as to moisture—*give no air at all.*

LASTREA.—*L. filix-mas,* the male fern, is the commonest (and some say the handsomest) species in Britain. It will grow anywhere and in any soil, but attains its fullest perfection in mellow loam or peat in a shady situation. The following varieties are fine—*cristata,* a magnificent object when well grown; *crispa,* a little gem for the case; *grandiceps,* a fine pot fern; *polydactyla,* a fine-crested variety.

L. æmula, the hay-scented fern, a charming species for the cool house or pot culture. When dried it is agreeably fragrant.

L. dilatata, the broad buckler fern, should be planted plentifully out of doors. The variety *dumetorum* has a beautiful rich appearance.

L. montana or *L. oreopteris,* the mountain buckler fern, common on Scottish moors, and by no means scarce in England and Wales. It should be planted out in loam and be freely supplied with water. It is not a good fern to grow in pots. When the hand is passed over the fronds a pleasant odour is emitted; when dried it has a sweet hay-like scent.

L. thelypteris, a free growing species for the cool house and for pot culture. It loves shade, moisture, and spongy peat, and travels fast.

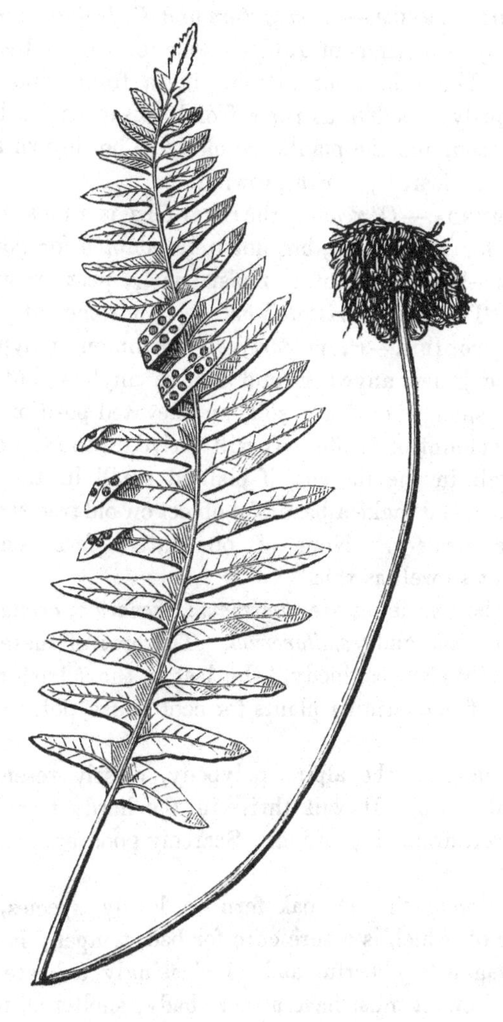

POLYPODIUM VULGARE

Ophioglossum.—*O. vulgatum* and *O. lusitanicum* are the only two kinds of Adder's-tongue fern in Britain. These should be kept in pots in a frame and conspicuously labelled, as their fronds disappear early in the season, and the plants are likely to be thrown away as dead. Scarcely worth growing.

Osmunda.—*O. regalis,* the royal fern, is a most noble plant for the garden, but not well adapted for pots or the fern-house. Plant in moist spongy peat or strong loam. The variety *cristata* makes a handsome pot plant.

Polypodium.—*P. vulgare,* the common polypody, may be grown anywhere and almost anyhow, but prefers a spongy or leafy soil, an elevated position and some amount of shade. It will thrive on the top of an old wall in the full sun if planted small in the first instance, and make a beautiful object on old tree stumps in the fernery. None of our native ferns endure drought so well as this.

The best varieties are *cambricum, crenatum, cristatum, omnilacerum,* and *semilacerum.* The first of these five is the "Welsh polypody," the last is the "Irish polypody;" five charming plants for cool house, pot, or case culture.

P. alpestre, the alpine polypody, closely resembles the lady fern. It will thrive in the hardy fernery if in a well-drained position. Scarcely good enough for pots.

P. dryopteris, the oak fern, a lovely species, the colour of which is a sure cure for bad temper; it is so extravagantly cheerful and so pleasingly delicate. If planted out it must have a very shady, sheltered, moist

POLYPODIUM VULGARE, *var.* CAMBRICUM.

place. It is one of the best ferns in the world for a ledge of rock in the cool fernery, or to grow in a large shallow pan as a specimen.

P. phegopteris, the beech fern; distinct and pretty, growing freely out of doors with the help of shade and moisture. A fine fern for pots and to plant near a fountain, as it attains its fullest beauty only in an atmosphere heavily charged with moisture. It must, however, be perfectly drained at the roots.

P. Robertianum, or *P. calcareum*, the limestone polypody, a pretty and peculiarly greyish-coloured plant which loves chalk or limestone rock, but will grow in almost any soil, and will endure the sunshine as patiently as *P. vulgare*.

POLYSTICHUM.—*P. angulare*, the soft, prickly shield fern, is the choicest of this section, a truly fine plant, sporting much and good in every form. It is so common that it will occur amongst the earliest "finds" of the fern hunter. It loves shade and a sandy, loamy soil, or leaf soil, but is not particular.

The following varieties are invaluable for pot culture, and the smallest of them well adapted for cases,—*concinnum, cristatum, grandiceps, grandidens, latipes, plumosum, proliferum.* The last named is a charming fern for pot culture, and thrives alike in frame, greenhouse, or stove.

P. aculeatum, the prickly shield fern, is at once distinct, bold, and handsome. Plant it in a shady spot and leave it alone for several years if you wish to see it thrive. A fine pot fern.

P. lonchitis, the holly fern, a handsome military-

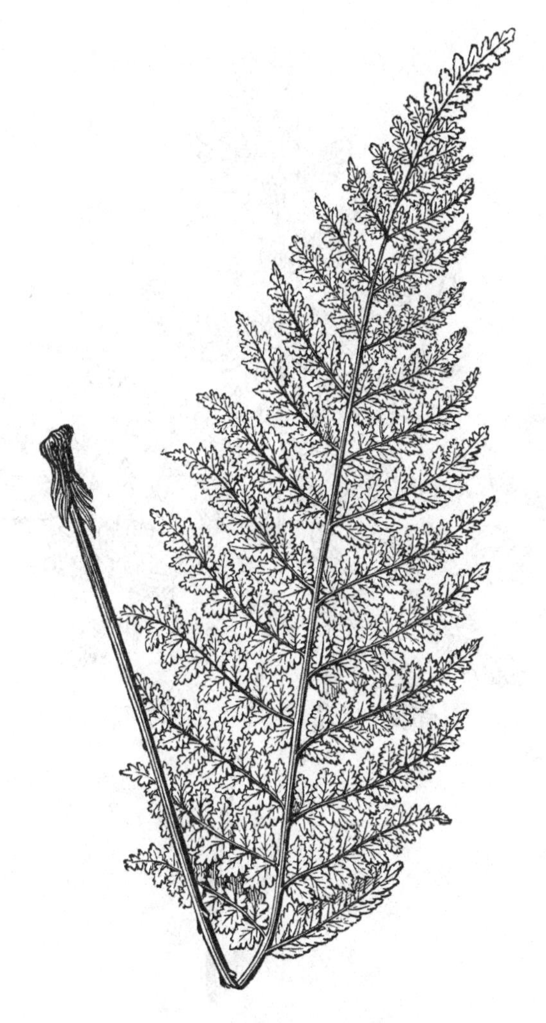

POLYPODIUM ALPESTRE.

POLYPODIUM DRYOPTERIS.

looking fern, rather difficult to manage, but deserving good generalship. If planted out give it a shaded, sheltered spot, and at least half a barrow full of a mixture consisting of loam two parts, peat one part, sharp grit and small broken bricks one part. It is a good pot plant if kept in a moist frame.

Pteris.—*P. aquilina,* the brakes, or bracken, is one of the best known of all. Plant it out in good loam or peat where it will have room to run, as it is a persistent traveller. Ten years ago I planted a piece not so big as my hand on a bank in my out-door fernery, and now it covers at least ten square yards of ground; at one point in its course it has crossed the gravel walk and come up on the other side. It makes a good pot plant, and also a good wall plant if planted at the foot of a shady wall and kept up by means of horizontally placed lengths of tarred string or copper wire. These supports should be placed about a foot apart; they will not be visible, and the effect will be a wall richly fringed as with climbing ferns. To see the bracken as it should be seen, we must go to the breezy moorland and skirt the warm woodside; it is, perhaps, the most truly rustic plant in Britain.

Scolopendrium.—*S. vulgare* is the common harts-tongue, one of the very first requisites of the hardy fernery. This plant will not live in the full sunshine, and it needs a good mellow loamy soil, or tough fibrous peat, with plenty of moisture to attain the growth it should, say a length of two to four feet. It is, however, an accommodating plant, as the fern hunter will soon learn by observation, for it will be found on damp

POLYSTICHUM ANGULARE.

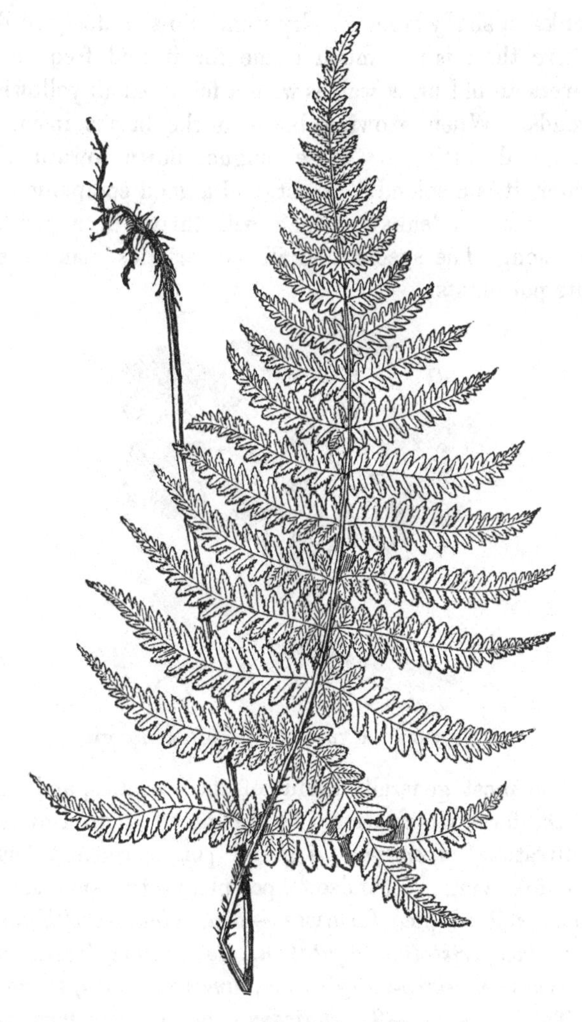

POLYPODIUM PHEGOPTERIS.

banks in shady lanes, on dry stone walls in dusty roads, where there is not much shade for it, and frequently covers an old brick wall as with a felt of small yellowish fronds. When growing between the bricks inside a well, and putting its huge tongues down towards the water, it is a splendid object, and a good companion to the true maidenhair, which will thrive in a similar position. The species and all the varieties make first-rate pot plants.

SCOLOPENDRIUM VULGARE RAMO-MARGINATUM.

The most generally useful of the varieties, and one of the ferns which should be first of all secured by the cultivation, is *crispum,* a grand pot or rockery fern. The following are handsome pot plants, the smallest of them well adapted for cases :—*bimarginato-multifidum, cornutum, cristatum, digitatum, glomeratum, laceratum, macrosorum, ramo-marginatum, ramosum-majus, Wardii.*

TRICHOMANES.—*T. radicans,* the Bristle fern, re-

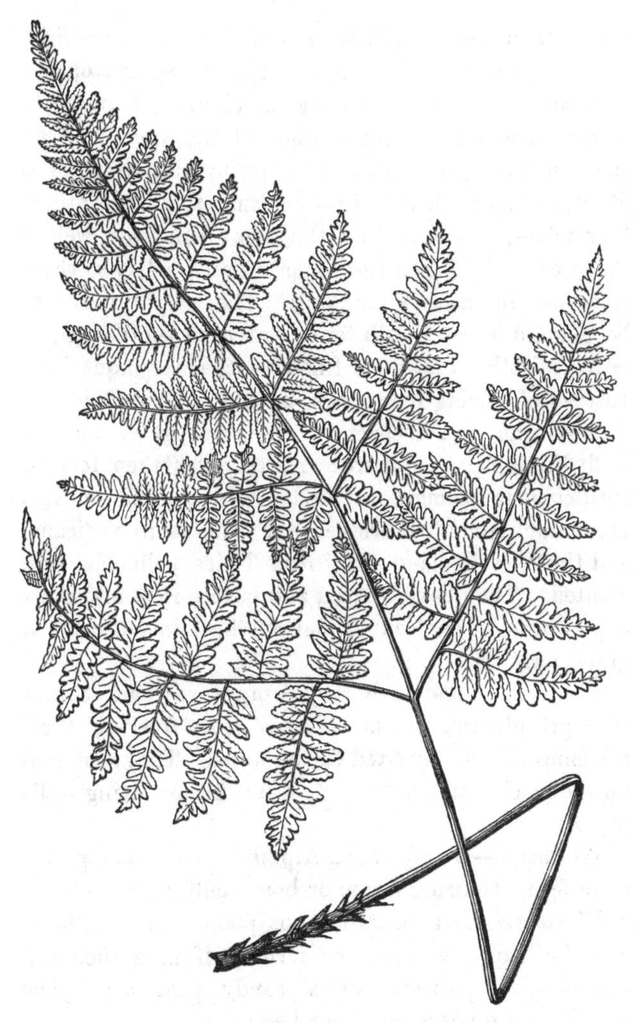

POLYPODIUM ROBERTIANUM.

quires treatment similar to that of the Tunbridge fern. As the roots are tough and wiry, and spread on the surface, it will be necessary in planting a piece to spread them out on the surface of the stone, and fix them in their places with pegs, or by placing nodules of stone upon them. In due time they will attach themselves, and after that the plant will grow well if taken care of. Small cases appropriated solely to these ferns are intensely interesting. They ought never to be planted in cases with ferns that need ventilation, as nearly all other kinds do. I had a large leaky aquarium. Instead of having it repaired, a hole was bored in the slate bottom, and a sheet of very stout glass was cut to fit the top. A miniature rockery was then formed with coke and cement in one large block, and on this Hymenophyllum Tunbridgense, Trichomanes radicans, and the New Zealand filmy fern Todea pellucida were planted. They have thriven and have a rich luxurious appearance. The leaky aquarium has thus become a grand fern case.

In a damp stone or a very damp warm dark corner of a greenhouse, the filmy ferns grow freely if their rhizomes are merely fixed to the walls. They soon run up the bricks, and form a delicate felt or living wall-paper.

WOODSIA.—*W alpina,* an Alpine Woodsia, is a pretty little fern, requiring frame or house culture.

W. ilvensis is a beautiful pot plant. It may, however, be grown in the open fernery, if in a sheltered well-drained position, in a sandy peat soil. The Woodsias are not adapted for beginners.

WOODSIA HYPERBOREA.

CHAPTER XII.

CULTIVATION OF GREENHOUSE AND STOVE FERNS.

PRACTICALLY the only difference in the management of the ferns of the greenhouse and the stove from those of the frame or cool fern-house consists in the increase of temperature proportioned to the character of the climates in which greenhouse and stove ferns are found growing wild. Various as are the climates and conditions in which ferns thrive on different parts of the earth's surface, they all become amenable to conditions nearly uniform when subjected to cultivation. Give the most delicate fern of the tropics treatment similar to what is advised for our native ferns, but with a higher temperature at every season of the year, and the chances are full ten to one that it will succeed perfectly. But undoubtedly it requires some judgment to assimilate conditions in the midst of which there occurs this important difference of temperature, and so we cannot expect to dispose of the subject of this chapter in any offhand or very general manner. However, we must beg the reader to recall the main points of our advice to this extent, that for outdoor, for frame, and for cool-house ferns, we have constantly recommended the use of a granular and mellow, loamy or peaty soil, a considerable degree of atmospheric humidity, shade from strong sunshine, and, in some cases, a very subdued daylight, as the conditions under which success is most likely to be secured. These several requisites are to be considered of the utmost importance in the cultivation of tender ferns, and the more so that the farther plants of any kind are removed from the circumstances natural to them, the more anxious should the cultivator be to provide for all their wants.

It is a common thing to see ferns and flowering plants mixed together in the same greenhouse or conservatory. It is quite possible to grow them well when so associated, but so few are equal to the task that when we meet with ferns and flowers in the same house, we usually find one or both in a deplorable condition of disease or imperfect development.

Ferns love shade and flowers love sunshine. Ferns thrive best in a still air, flowers usually require a moving atmosphere, and many kinds that are most highly prized need abundant ventilation. As to atmospheric humidity, while ferns with very few exceptions enjoy abundance of it, there are not many kinds of flowers capable of enduring without injury the degree of aerial moisture that would benefit the growth of ferns. These are important considerations which we are bound to place before the reader at this juncture, for indiscriminate associations of plants in stoves and greenhouses are the causes of many and bitter disappointments. While this matter is before us, however, it should be said that if due care be exercised, many kinds of flowering plants may be grown in the same houses with ferns, if the selection is made judiciously in the first instance, and the best positions as to air, light, &c., are selected for them. Thus, as to sorts it will be found that camellias, azaleas, cyclamens, primulas, liliums, oleas, and statices, are well adapted to associate with greenhouse ferns, if the sunniest positions are assigned them; on the other hand, heaths, pelargoniums, echeverias, epiphyllums, boronias, epacris, and kalosanthes, are far less suitable, needing more air and sunshine than most ferns could endure without injury. It must be remembered, however, that many beautiful plants, such as palms, for example, may be grown with ferns to afford variety, and the same routine of treatment will suit both. In the stove it is common enough to find achimenes, gloxinias, alocascias, caladiums, begonias, gesneras, and marantas, associated with ferns without the least injury to either. Yet in the full blaze of sunshine, where a croton or an ixora would thrive, it would be almost impossible for a fern to live, except in the form of a disgrace to its possessor. So far we see that compromises are possible. There is yet another mode of associating ferns and flowering plants in the same house, and that is to make banks and rockeries beneath the stages where shade and humidity will favour the growth of ferns, and render positions otherwise useless and unsightly as attractive nearly as the stages themselves, on which the amaryllids or the pelargoniums are blooming bravely. A bank of peat faced with large burrs answers admirably for a fernery of this sort, and the varieties of cystopteris, woodsia, scolopendrium, and selaginella, are pretty sure to take to it readily, while in the most select spots, hymenophyllums, trichomanes, todeas, and maidenhairs, will soon become established, and acquire a luxuriance of growth without the least care, such as to make a mere mockery of all our closed cases and bell-glasses, and curious caves constructed expressly for the cultivation of these gems of the fern garden.

There cannot be a doubt that the plan recommended in Chapter VII for the cultivation of hardy ferns under glass is the best also for greenhouse and stove ferns, unless it be the desire of the cultivator to have the whole or a part of the collection in pots, in which case, of course, something in the nature of a stage or table becomes necessary. A spacious fernery adapted for ferns of all climates, and for the display of them under circumstances which we may justly describe as *natural*, forms one of the most valuable embellishments a garden can boast—enjoyable at all seasons, and especially so in winter, when rough weather forbids our seeking open-air enjoyments, and when, perhaps, if weather permitted, we should find but little in the garden or the field to interest us. One of the best structures of the kind I am acquainted with is in the garden of Alfred Smee, Esq., Carshalton. The walls are formed of solid banks of peat, which extend on either side of the plate on which the rafters rest, so as to form borders within and without. The house may be about eighty feet in length, the banks on either side are varied in outline, and there is in one spot a basin tenanted with gold fish, and surrounded with ferns of peculiarly novel aspect, which are constantly bedewed by the spray from a fountain. The roof is a span running east and west; the south side of it is covered with felt, and the north side with glass, a plan which admits abundance of light, and renders shading wholly unnecessary. The whole structure is placed on a slope, the lower part being considerably below the outside ground level. At this lowest part is placed the furnace, and there is an extra service of pipes there to maintain a stove temperature. In the middle of the house there are fewer pipes, and a greenhouse temperature is kept. At the upper end the pipes suffice only to keep frost out. Thus in one house the ferns of tropical, temperate, and frigid zones are all accommodated, and though the whole structure is rough, and has been constructed on the most economical principles, the interior presents at all seasons a grand spectacle, and affords a most delightful promenade.

Although, as explained above, ferns and flowering plants may be grown together, those who would do justice to the former must appropriate a house to them exclusively. It is possible to adapt a south aspect to the purpose, but it is not advisable to encounter such a difficulty. A north or north-west aspect is the best. The house should have a roof of not very steep pitch, a sufficient service of hot-water pipes, and ventilators near the pipes to afford warmth to the fresh air as it enters, and others in the roof at each end, but none elsewhere unless the house is a large one. A frequent change of air is essential to the health of the ferns, but we do not want a rushing wind or so much ventilation as to render the air of the house so dry that the fronds will lose their freshness and health.

Thousands of villas are now furnished with what are called "conservatories," which would answer admirably for ferneries where they happen not to be exposed to burning sunshine all the summer long. The sunniest of these little glass annexes answer admirably for grape vines and succulent plants, such as cactuses and echeverias; the shady ones would answer admirably for ferns, whether in pots or planted out in miniature rockeries.

In the management of greenhouse and stove ferns the most important matter is to secure a suitable temperature for each department or group of plants. The greenhouse kinds require a temperature of 40° to 50° all the winter, but from the middle of April until the middle of October artificial heat may be dispensed with altogether, unless the weather is exceptionally cold; and stove ferns require a temperature ranging from 60° in winter to 90° in summer.

In every case the amount of moisture must be proportioned to the temperature, the more heat the more water, both above and below. When the plants are growing freely the syringe should be used to produce a fine shower over them once or twice a day, and water should be sprinkled on the floor to cause an abundant evaporation. They will also require plentiful supplies of water at the roots.

There is no large class of plants in cultivation for which we may so safely give general cultural directions as for ferns, yet certain kinds require exceptional treatment both in heated houses as in cool ferneries and the open air. The gold and silver ferns, such as gymnogrammas, are for the most part highly susceptible of injury through excess of water, especially when administered by means of the syringe. All the tree ferns such as Dicksonias require abundant supplies of water, especially over their ample fronds. Most of the kinds which have thick succulent leaves, such as Niphobolus, require drier positions if planted out, and extra careful drainage if in pots, than others that are of flimsy texture.

The cultivator must be careful to regulate heat and moisture in such a manner as to ensure to the plants regularly recurring seasons of activity and rest. When new growth commences in spring there should be a gradual augmentation of temperature and humidity to afford needful stimulus and support. When in autumn growth should naturally cease, the supplies of heat and moisture should be diminished; and during the winter rest should be promoted by keeping the house as cool and dry as is consistent with safety. It is bad policy to expose ferns to hardships, such as deferring the lighting of a fire until the fronds are actually frozen or mildew has marred their beauty, for the next season's growth is jeopardised by such treatment, and some valuable plants may be lost entirely. At the same time the cultivator may take comfortable assurance from the fact that the majority of this class of plants are exceedingly accommodating; they will at times bear without material injury more damp, more draught, more sunshine, and lower degrees of temperature than a prudent adviser on their cultivation would dare to recommend as good for them. Fully half of the whole number of stove ferns known to cultivators have been well grown in greenhouse temperature, and a very large proportion of greenhouse ferns, properly so called, have been grown to perfection, without any aid from artificial heat, in our own garden. Our large specimens of Adiantum cuneatum, Asplenium biformis, Blechnum brasiliense, &c. &c., that we have exhibited in public, have never known a taste of artificial heat from the time when they started from spores under bell-glasses until they attained their present dimensions of a yard or so across. This adaptability is particularly exemplified in the cultivation of ferns in closed cases, Mrs. Hibberd's cases containing delicate ferns of the tropics side by side with the natives of the British woods, yet all in the most perfect health and beauty.

The soil for pot ferns should always consist in great part of vegetable mould and sand; mellow loam, silky to the touch and crumbling to powder between the fingers without soiling them; peat of a brownish rather than a blackish cast, and containing an abundance of vegetable fibre, so as rather to require tearing than crumbling to reduce it; sand of a sharp clean nature: these three ingredients are sufficient for the preparation of a universal fern compost. In the case of very small delicate habited ferns use two parts peat, removing all the rougher portions, and one third sand. For full growing and rather large plants use two parts peat, one part loam, and one part sand, the rougher fibrous portions to be laid over the crocks, and the bulk to be used in a rather lumpy state. For very robust habited sorts of large growth the compost should consist of two parts loam, one part peat, and one of sand, with a liberal addition all through of broken brick or tile of the size of walnuts or hazel-nuts. Ferns that require a drier soil than ordinary should have a compost containing more sand, less loam, and the addition of a considerable proportion of pounded bricks or charcoal.

No particular kind of pots is necessary for the cultivation of stove and greenhouse ferns, but, as a rule, they do not root deeply, and shallow pots are to be preferred. Those we use for specimens are made for us by Messrs. Adams, of the Potteries, Belle Isle, King's Cross; they are extra stout in substance, carefully finished, and well burnt, and in proportions wider than their depth. A favorite size with us for medium specimens is thirteen inches wide (inside), and nine inches deep. In these we allow our specimens to remain two or three, and even four or five years, without being repotted, though, as a rule, all pot ferns should be repotted annually in February or March, both to repair the defects of the drainage and remove effete soil, and supply fresh food for maintaining a vigorous growth.

In every case thorough drainage is of the utmost importance, and no progress can be made in fern culture unless the operator pays especial attention to this matter. As for whatever else may be requisite to crown your labours with success, I will endeavour in what follows to indicate as clearly as I can, but it is very certain I shall leave unsaid much that might be said, but I may, even thus far, have assisted you to read the Book of Nature to advantage, so that at the point where I stop your studies will take a better direction under authority which never fails.

> "Our needful knowledge, like our needful food,
> Unhedg'd lies open in life's common field."

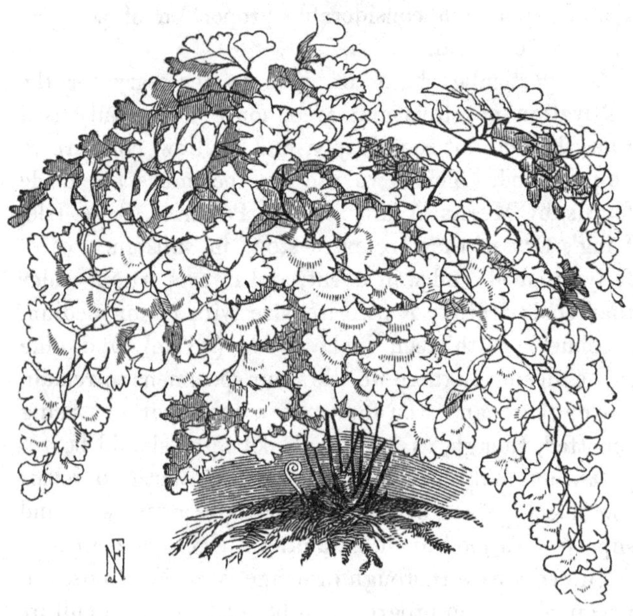

ADIANTUM FARLEYENSE.

CHAPTER XIII.

FIFTY SELECT GREENHOUSE FERNS.

THE selections I shall make in this and the next chapter will comprise ferns of the most distinct and various characters, essential in any collection in which beauty and character are the qualities most desired, and all of them suitable for beginners in cultivation. None of the gold and silver ferns will be included in these selections; they will be dealt with separately, as needing more skill and care than beginners are likely to bestow upon them. Technical descriptions are not to be thought of in a work of this kind.

Anemidictyon phyllitidis, a pretty flowering fern, adapted for pot culture, or to plant out, or for the fern case.

Adiantum assimile, A. cuneatum, A. formosum, A. fulvum, A. tinctum, a charming group, requiring shade, not rooting deep, and better if they never have water over their fronds. They are all adapted for specimen culture, the last is one of the most elegant in form, and has a rosy purplish tint on its young fronds.

Asplenium bulbiferum, A. caudatum, A. dimidiatum, A. dimorphum, A. hemionitis (or *palmata*), *A. lucidum,*

ADIANTUM CUNEATUM.

A. obtusatum, A. præmorsum. The two most striking of this group are hemionitis and dimorphum, which should be first secured. As to management, the merest beginner can grow them well.

Blechnum occidentale, B. brasiliense, two noble ferns, suitable for the greenhouse, yet rather tender, and utterly incapable of bearing a touch of frost.

Campyloneurum phyllitidis, a very distinct entire fronded fern, which forms a striking object when well grown. It is commonly kept in the stove, but the greenhouse is the proper place for it. The soil for this fern should be rich and gritty, containing plenty of fibre, but it should not be deep, as it is a shallow rooter. Abundance of water should be given while the plant is growing. It is not particular whether in sun or shade, but, of course, will not bear roasting.

Davallia canariense, the "Hare's-foot" fern, *D. dissecta,* a charming pair, and the easiest of the family to grow. It is easy to kill Davallias by means of heavy soil and excess of water; equally easy to grow them to perfection with plenty of drainage, a very gritty soil, and water in moderation. The fleshy rhizomes must be pegged out upon the surface in planting new pieces.

Gleichenia flabellata is the only one of the genus I can recommend to a beginner. It is a fern of large growth, requiring to be carefully trained like a delicate climbing plant. Plant in a shallow pot, give plenty of water and plenty of air. When you have mastered this one add *G. dicarpa* and *G. speluncæ.*

Goniophlebium appendiculatum, a splendid edition (we

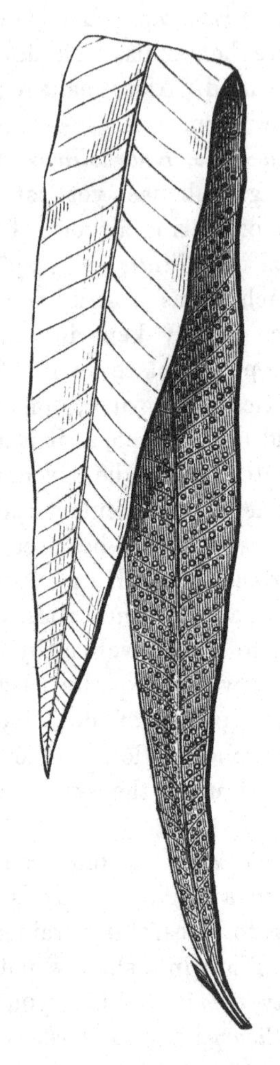

CAMPYLONEURUM PHYLLITIDIS.

may call it) of our own common polypody; when young tinted with crimson. It requires a rather dry

DAVALLIA CANARIENSIS.—THE HARE'S-FOOT FERN.

soil; so add extra sand and a good sprinkling of fine

potsherds to the compost. Do not wet the fronds at all.

Hypolepis tenuifolia, a finely divided brightly coloured fern, requiring abundance of water.

Litobrochia incisa, rather coarse, but worth having; it will take care of itself almost anywhere.

Lastrea quinqangularis, L. patens, two exquisite gems, cheap, but not common. They thrive in our cool fernery.

Lomaria magellanica, L. gibba, L. chiliense, grand ferns, nearly hardy, and indispensable in even the smallest collection. L. gibba will endure almost any hardship except frost.

Lygodium Japonicum, L. scandens, the two best "climbing ferns" for a beginner. They may be trained to sticks or wires in the same way as a convolvulus.

Mohria thurifraga, a rich fern, good enough for exhibition. It thrives in the cool fernery, but is rather tender.

Nephrodium molle corymbiferum, a charming tasselled fern, like a cockscomb; rather tender, and therefore pretty sure to be lost if kept damp and cool in winter; yet it is a greenhouse fern, and one of the best.

Niphobolus lingua, N. rupestris, pretty entire-fronded ferns, requiring a dry soil, with plenty of broken brick and sand. Well adapted also for the fern case. For several years past we have grown a collection of ferns of this class in a sunny part of the geranium house, and the full blaze of the sun has agreed with them perfectly. Any excess of moisture will kill them.

Nephrolepis tuberosa is the only one of this splendid family I can recommend for the greenhouse, though they are all classed as greenhouse ferns in trade catalogues. This, however, is so distinct, that you must have it if you buy only a dozen.

Onychium Japonicum, a delicate fennel-like fern,

NEPHRODIUM MOLLE, *var.* CORYMBIFERUM.

fragile, fairy like, yet nearly hardy, and always in health, if thoroughly shaded.

Platyloma rotundifolia, very distinct and fine when in fruit. It must have deep shade.

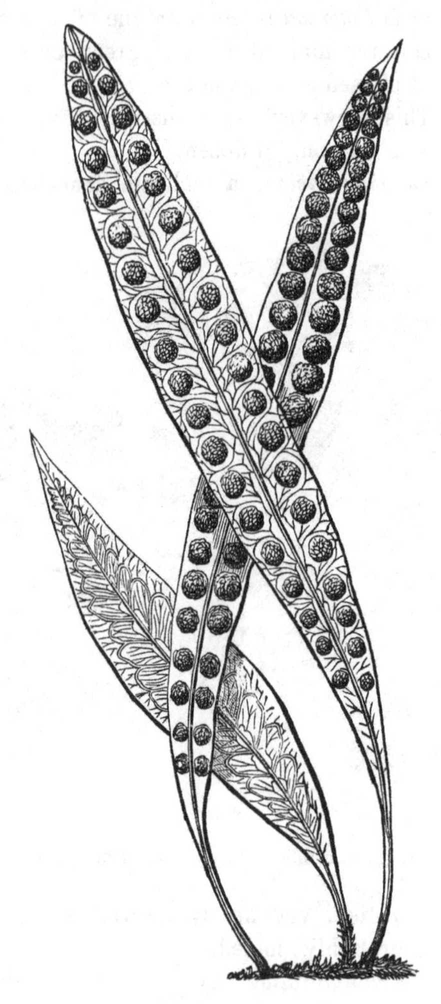

POLYPODIUM VENOSUM.

Polypodium venosum in the way of Niphobolus lingua, a charming object when its ruddy fruits are ripe. This fern requires peculiar treatment, and if properly planted in the first instance will occasion no trouble whatever. In any case the roots must be extra well drained, for stagnant moisture is certain death to this plant. The soil which suits it best is a mixture of equal parts gritty leaf-mould, sandy peat, and potsherds broken to the size of peas. In such a mixture, not more than six inches in depth (four inches is sufficient), on a bottom of some material which will allow of ready escape for surplus moisture, the plant will do well, and prove itself an almost hardy fern. Obviously the best way to deal with a plant so constituted is to suspend it. When grown in a basket in a warm greenhouse it soon forms a fine specimen, the tawny rhizome creeps about wildly, and soon covers the basket with a beautiful complexity of cord-like windings, and from every part of it, except the young pushing shoots of the season, barren and fertile fronds are produced in plenty. To increase it is easy enough; cut off a portion of rhizome with fronds and roots attached; pot it in the same sort of mixture as is recommended for specimen plants, and give it proper encouragement, and it will soon make a plant.

Phlebodium aureum, P. sporodocarpum, two bold glaucous tinted ferns, with ruddy rhizomes that run upon the surface. They are both classed as stove ferns in the books, but they are as easy to grow in a greenhouse as any in this list; at all events we can keep them in luxuriant condition in the cool house. Plenty of grit in the soil, and perfect drainage.

Polystichum setosum, a lovely dark green fern, will take care of itself anywhere in the shade.

Pteris ternifolia, P. hastata, P. cretia albo lineata, P. scaberula, P. flabellata, a fine group; scaberula runs about, and should not be put into a case for that reason; but in a basket, which will allow it to peep out, it is at home; as for the last in the list it is lovely, and thrives in our cool house.

Platycerium alcicorne is absolutely indispensable for its curious habit and its hardiness. It will bear seven or even ten degrees of frost, and yet come right again, but should never be so much punished. Get a block of old wood, scoop out a hole, and put in it some fine peat, and in that hole fix the plant firmly. Then hang up the block by means of copper wire, and syringe frequently all the year round. It will in time cover the block with its tawny shields (we call them "pot lids"), and make a grand object. A plant has hung near the roof of our cool house for ten years, and has several times been frozen.

Todea pellucida and *T. superba* are a pair of New Zealand filmy ferns of the most exquisite character. I am half afraid to recommend them to beginners, yet they only want deep shade and moisture to succeed to perfection, as they are nearly hardy. Plenty of drainage, plenty of patience, as little air as possible, and all will be well. I have some fine plants growing in a disused (because leaky) aquarium; they are in fine condition. They are covered close with a sheet of glass and never have any air at all.

Thamnopteris australasica is too good to be omitted.

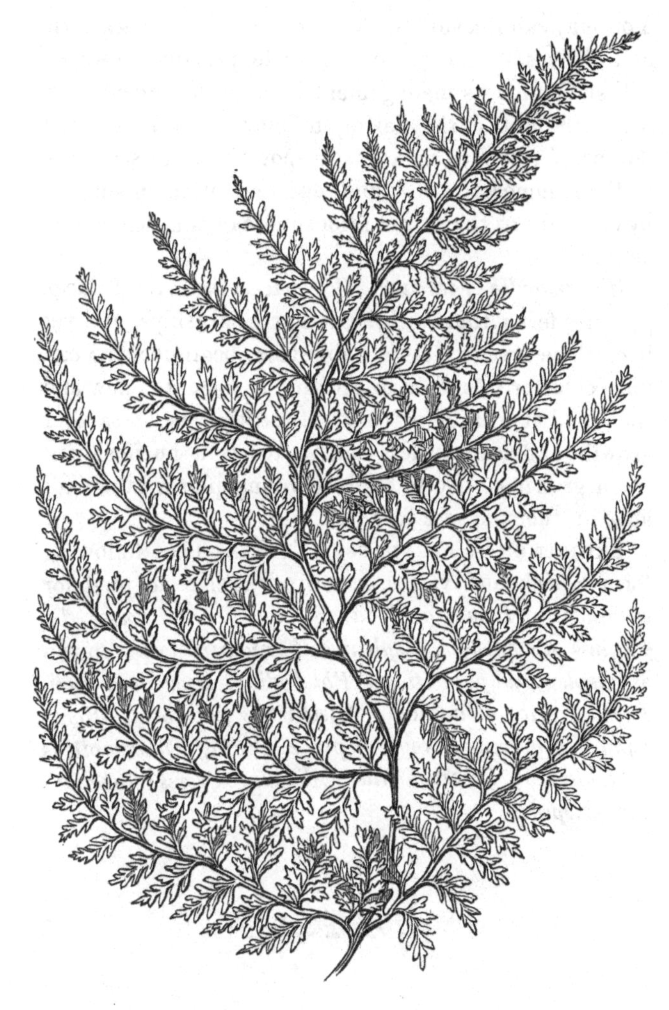

PTERIS SCABERULA.

You may call it a sublime hartstongue. It loves warmth, and thrives in the stove. A little practice, however, will suffice for its management in a warm greenhouse. Mr. Gibson had the daring to make a bed of a few dozens of this fern in a shady spot in Battersea Park in the summer of 1867, and not one of them suffered by exposure to the vulgar atmosphere of this degenerate clime.

Woodwardia radicans, W. orientalis, grand large growing ferns that will bear many hardships, and yet live. The first is indispensable to a beginner who can find room for it, and as to growing it, look at it now and then, and it will be satisfied; the other is of smaller growth, and scarcely less hardy; it has a purplish tint when growing. Both produce young plants in abundance on their mature fronds.

Exhibition Greenhouse Ferns.—The following form a rich and varied group of twelve adapted for exhibition: *Lomaria gibba, Blechnum brasiliense, Asplenium dimorphum, Asplenium hemionitis* (also known as *Asplenium palmatum*), *Phlebodium sporodocarpum, Pteris cretica albo-lineata, Gleichenia flabellata, Microlepia platyphylla, Nephrolepis exaltata, Thamnopteris australasica, Woodwardia radicans, Pteris flabellata* var. *crispa.*

CHAPTER XIV.

THIRTY SELECT STOVE FERNS.

NEIMIA collina, a fine representative of an interesting group of flowering ferns. It requires the most commonplace treatment.

Adiantopsis radiata, a very distinct and elegant little fern; the divisions of the fronds radiate in a regular manner from a common centre.

Adiantum concinnum, A. Farleyense, A. macrophyllum, A. tenerum, A. trapeziforme; a splendid group, not one of which can be dispensed with in even the smallest collection. A. Farleyense might in an offhand way be pronounced the most beautiful fern known, but the assertion would not bear criticism, just because there are so many beauties of the kind; it is impossible to decide which is the best amongst them.

Asplenium formosum, A. serra, A. viviparum. The second of these is a large-growing exhibition fern; the other two are delicate beauties.

Blechnum brasiliense, a noble fern, well adapted for exhibition, and one of the easiest to manage.

Brainea insignis, a grand fern, palm-like in growth, the young fronds tinged with a lovely rosy hue.

Davallia polyantha, D. aculeata. The fronds of the

first have a rich rosy crimson tinge when young; the

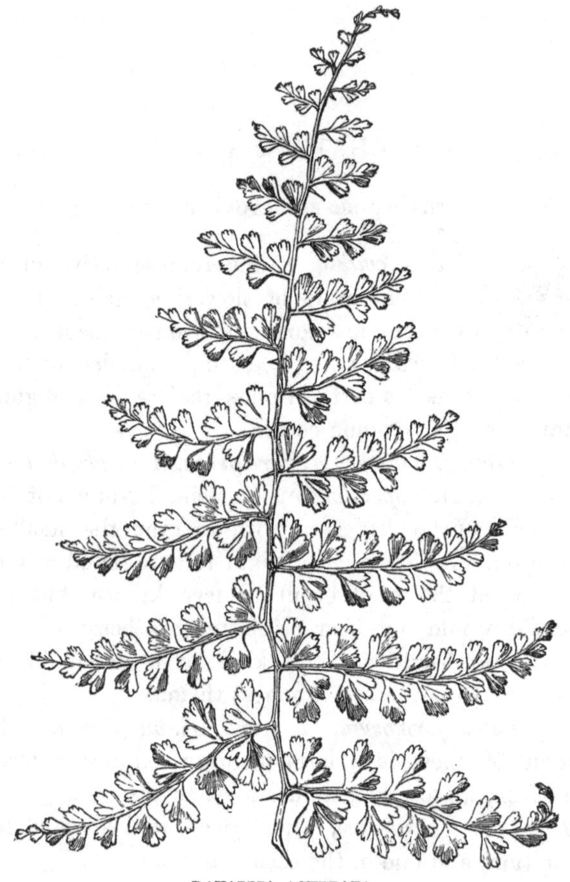

DAVALLIA ACULEATA.

other is as thorny as a bramble, and grows in the style of a climbing fern.

POLYPODIUM LACHNOPODIUM.

Elaphoglossum frigidum, a curious and most beautiful species, with entire wavy, pendant fronds, which are covered with grey scales, giving it a hoary appearance. Nothing in its way can surpass it.

Gleichenia pubescens, one of the finest and easiest stove plants of this section. Deserves all the care that can be given it to form a fine specimen.

Goniophlebium fraxinifolium, a particularly handsome once divided fern, of a delicate pale green colour.

Goniopteris crenata, extremely pretty when in fruit, and well worth growing as a specimen.

Hymenodium crinitum, most distinct and beautiful; not in the least resembling any other fern known; the fronds are like the large leaves of some tropical tree, densely bearded with black hairs.

Hemionitis palmata, a distinct ivy-like fern, bearing many tiny young plants on its fronds.

Lomaria attenuata, a very pretty little blechnum-like fern, the young fronds of which have a delicate rosy hue.

Lygodium flexuosum, the grandest of the climbing ferns; scarce.

Nephrodium glandulosum, extremely pretty and peculiar; the fronds once divided; shining green.

Nephrolepis exaltata, N. pectinata, the two best table and sideboard ferns known, and first rate, too, for the centre of a fine vase or large case group. We have lost many fine plants of both species in the endeavour to make greenhouse ferns of them.

Polypodium lachnopodum, P. Henchmanni, P. phymatodes, three fine and very distinct species; the metallic blue colour of the second is peculiar and pleasing.

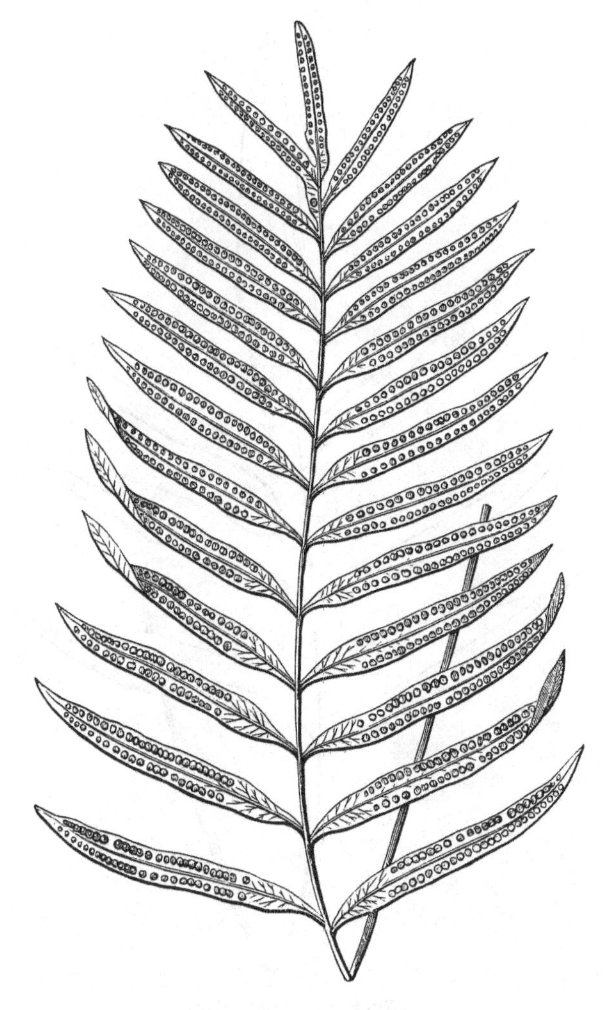

POLYPODIUM HENCHMANNII.

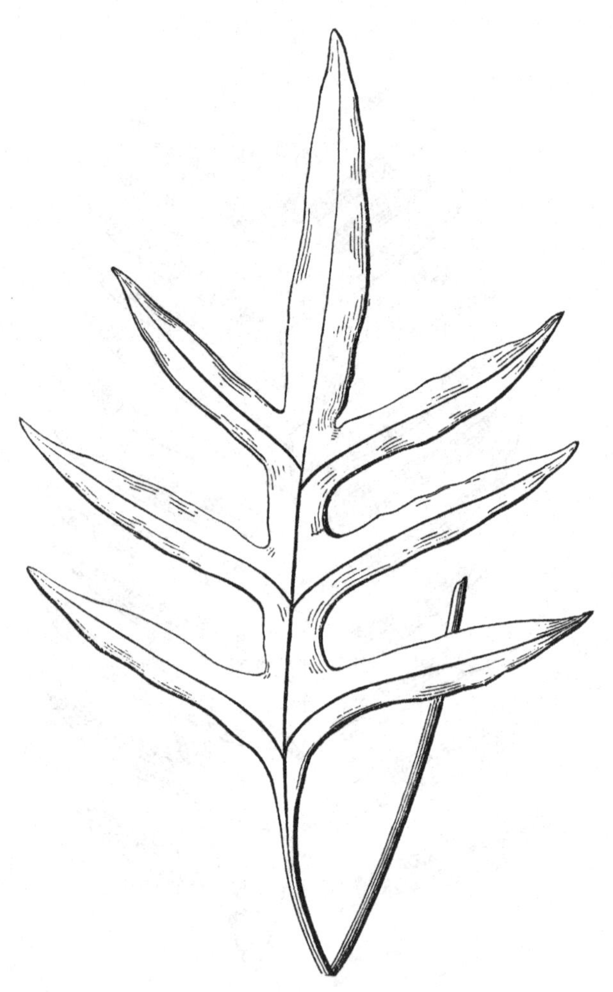

POLYPODIUM PHYMATODES.

Pleopeltis membranacea, a scarcely interesting fern at first, but one likely to become a special pet in time. It dies down completely in winter, and comes up again in the spring. The fronds are undivided, and bear a remote resemblance to lettuce leaves. We have had some plants five or six years in an unheated case, but it is delicate, and most at home in the stove.

Pteris argyrea, P. aspericaulis, two richly variegated ferns, which are very subject to attacks of thrips if kept in a dry air. *P. tricolor* is a favourite which I do not recommend because troublesome to grow, and scarcely worth growing.

Platycerium grande is the finest of the stag's horn ferns, and though usually described as a greenhouse plant, attains a far finer development in the stove. Fix it on a block of wood, and suspend it, or put a block in a pot, and place the plant near it, so that it can take hold and cover the block in its own way.

EXHIBITION STOVE FERNS.—The following form a rich and varied group of twelve adapted for exhibition: *Adiantum Farleyense, Adiantum trapeziforme, Hymenodium crinitum, Aspidium macrophyllum* (also known as *Cardiochlæna macrophylla*), *Asplenium myriophyllum* (also known as *Asplenium cicutarium*), *Asplenium serra, Drynaria morbillosa, Gleichenia dichotoma, Lygodium flexuosum, Nephrolepis davallioides, Platycerium grande,* and *Pteris argyrea.*

CHAPTER XV.

GOLD AND SILVER FERNS.

NONE of the so-called gold and silver ferns are adapted for beginners. They are so superbly beautiful that people altogether unaccustomed to ferns buy them and put them in greenhouses, supposing that watering now and then is all the care they want, and in the course of a month or so the plants die, and an absurd inference is drawn from the occurrence that ferns in general are impossible things. It is quite certain that a very large number of maidenhair ferns are killed by ladies who pretend to love ferns and really have no real care for them at all; but probably there are more gymnogrammas killed through absurd treatment than any other class of ferns whatever. Yet they require but little more care than most others; their peculiarity is that if that care is denied them they die outright; whereas many other kinds survive neglect and ill-treatment, and regain their cheerful looks "in no time" if proper treatment is resorted to.

If we could repeat in an intensified form all the cautions that have been given in this work up to this point we should have a practical code for the cultivation of gold and silver ferns. Instead of attempting that, I will sketch out a code in a very few words, begging the reader to regard each word as pregnant with meaning, each hint and direction as involving for the ferns issues of life or death, as they may be observed, trifled with, or ignored. The pots must never be larger than the plants can soon fill with roots. They must be very carefully drained by means of potsherds packed with the greatest care. The soil should consist of good fibrous peat and a large proportion of sharp siliceous grit; silver sand is almost too fine, but must be used if nothing more granular is obtainable. The plants must be potted firmly with the crowns well above the surface. Thenceforward the temperature and the degree of humidity are of the utmost importance. Only a small proportion of all the gold and silver ferns in cultivation require the heat of the stove, but not one of them will endure a lower temperature than that of the house it properly belongs to. Thus, there are many stove ferns that thrive in a greenhouse, and many greenhouse kinds that do well in an unheated house. But it is not so with those before us; they are not accommodating, they are exacting, and must be humoured to their whim. As to moisture none of them will bear much; to make them very wet is to put them in jeopardy. But on the other hand to let them go dry is certain death. The principal enemies that make war against them as cultivated plants are imperfect drainage, heavy soil, cold, damp, and drought. In no case should the fronds be wetted by the use of the syringe. The little that I have said compasses the whole subject, and the observant cultivator, who is also diligent and constant in his work, will find that the secret of success with this class of ferns is *unremitting attention.*

The following are the names of the best gold and silver ferns in cultivation:

Adiantum sulphureum, the Golden Maidenhair. This exquisite plant only needs careful greenhouse treatment.

Cheilanthes argentea, a delicate silver fern; greenhouse. *C. borsigiana,* golden; stove. *C. farinosa,* silver; a fine species very distinct, requiring great care; stove. *C. fragrans,* a lovely little gem tinged with orange, well adapted for greenhouse or case. When dried agreeably fragrant. *C. elegans,* silvery, a most delicate and much prized fern, best grown in a warm greenhouse, in a compost of lumpy peat and broken bricks or stone. *C. pulveracea,* the under side silvery, the edges golden: a fine companion to *C. farinosa,* and needing the same treatment.

Gymnogramma chrysophylla, the finest of all gold ferns; it must be grown in the stove. *G. Peruviana argyrophylla,* silvery-grey on both surfaces, a splendid stove fern. *G. ochracea,* slightly golden, easy to grow, but needing to be kept in the stove all winter. *G. sulphurea,* a pretty little plant, light green above, sulphur-yellow beneath; must have stove treatment. *G. Tartarea,* the under side of the fronds pure silvery-white, the best of all silver ferns for beginners; it thrives in the stove, but may be kept in good condition in a greenhouse.

Nothochlæna argentea, a fine silvery companion to *Cheilanthes farinosa,* and requiring similar care. *N. flavens,* an exquisitely beautiful miniature golden fern;

CHEILANTHES ARGENTEA.

a good companion for *N. nivea,* which is equally diminutive and densely powdered with silvery farina.

GONIOPTERIS CRENATA.

CHAPTER XVI.

TREE FERNS.

TREE ferns have been brought within the reach of fern growers who happen not to be millionaires, by the enterprise of trade collectors, and may be purchased according to size, rarity, &c., at from five guineas each and onwards. Those, however, who would like to grow their own, and who are blest with the needful patience, may obtain young plants to begin with at from five to fifty shillings each. There is much to be said in favour of purchasing young plants; they are extremely ornamental, and the greenhouse kinds will thrive in the shady parts of a conservatory where scarcely anything else would grow. If it is intended to embark in the purchase of fine specimen tree ferns it will be important to consider first the space available, for the spread of a fine Dicksonia or Cyathea is considerable, and it is not good for them to rub their fronds against the glass roof, however carefully it may be shaded.

There are no species of filices more easy to cultivate than such as are classed as "tree ferns." The soil should be the best peat in a rough state, with but little sand added; the addition, however, of sphagnum moss or cocoa-nut fibre improves the peat for the purpose. Large pots or tubs are needful; the roots will bear a certain amount of cramping, but as a free growth is desirable—in fact essential—both to maintain the health besides developing the beauty of the plants, as much pot room must be allowed as possible, consistent with the sizes of the plants and the place they are kept in. Shade is of the first importance, abundance of moisture is indispensable.

The most desirable greenhouse tree ferns are *Dicksonia antarctica, D. squarrosa, Alsophila australis, A. excelsa, Cyathea dealbata.* The first named is the most useful and is extremely likely to prove a hardy plant for sheltered shady dells in the south-western parts of England and the warmer parts of Ireland. The beginner should avoid *Alsophila capensis* as risky, and the expert need be in no hurry to obtain it.

The most desirable tree ferns for the stove are *Alsophila glauca, A. armata, Cibotium scheidei, Cyathea arborea, C. microlepis.*

Let us now suppose that some obliging friend in Australia makes you a present of a lot of tree ferns. He has found some specimens with stems from four to five, or even six feet long; he has cut away all the fronds, and dug them up, without taking the trouble of saving any of the roots. In fact, they are stems and nothing more—stems, sans fronds, sans roots, sans everything. He leaves them out in the air for a few days to dry, and then packs them with shavings in a box; let him be especially careful that this box be not air-tight—that is their greatest danger. In this way they generally come with pretty good success, a large majority of them quite safely. And now, as we unpack them, let them be placed upright in some close, cool, dark corner—under the stage of a greenhouse is as good a place as they can have. Give them a syringing once a day for the first week, and after that two or three times a day; never allow them to get quite dry. By the end of a fortnight, or even sooner, you will observe the points of new roots starting out upon the stem, and the closely coiled-up fronds in the centre to be pushing upwards.

They may now be safely potted. I have no faith in *exact proportions* for mixing soils, and my candid opinion is that the mechanical condition of the soil has more influence than anything else. Let it then, above all things, be open and porous. Use pots as small as you can in the first place, and shift them from time to time as the plants may require it, using rough peaty soil as before. If allowed to become pot-bound, the fronds soon dwindle in size. Keep them always moist at the root, and during nine months of the year the stem should be kept constantly moist. This can easily be done without wetting the fronds much, which is not always beneficial. Do not expose your plants to draughts of dry air, and be sure to shade them from bright sunshine. Following these simple rules, your tree-ferns will be an ever-increasing source of pleasure.

CHAPTER XVII.

FERN ALLIES.

LYCOPODIUMS, SELAGINELLAS, PEPPER-WORTS, HORSE-TAILS, AND MOSSES.

HOWEVER slightly the cultivator of ferns may be interested in their technical classification and botanical affinities, it is impossible to proceed far in the practice without being attracted by the beauties of certain plants which are not ferns, but cousins-german to them. A tuft of club moss in a marsh or of horsetail in a damp and tangled hedgerow will arrest attention, and the beginner may ask the question—"Is this a fern?" and when assured that it is not one, will probably ask again, "What is it?" In the greenhouse and the stove the moss-like selaginellas, usually called lycopodiums, associate with ferns as their proper companions, requiring similar treatment and being obviously allied in structure and habits. On this border land there is entertainment for the curious; a knowledge of the distinctive character of the tribes of plants that haunt it will prove, like many other kinds of knowledge, abundantly remunerative to those who will earnestly seek it, and the lover of vegetable beauty, who may be indisposed to pore over books or labour with the microscope will discover here many gratifications.

The true ferns may be traced through many gradations of physiological structure with comparatively little trouble. At all events when the botanists have classified them, it needs no subtlety of perception to determine that the adder's tongue and moonwort ferns are the lowest in the scale, and that their very existence is suggestive of a gradation of similar forms laterally or vertically separated from them to which these least fern-like ferns serve as connecting links. The plants that are closest allied to the ferns are the Lycopodiums, the Selaginellas, the Pepperworts, the Horsetails, and the Mosses. After these we get amongst lichens and fungi, and as we must stop somewhere, the foregoing five families are all we shall recognise for the purposes of this chapter. Each family contributes beautiful plants adapted for the fern garden, and as for the selaginellas they are all beautiful, and we make selections from amongst them, because usually we cannot find room for the fifty or more species and varieties known to cultivators.

Lycopodiums and Selaginellas closely resemble mosses in their branches and leaves, while in many of their general characters and aspects they bear close resemblances to ferns. They are, however, distinct from either, and are especially characterised by the nature of their leaves and their fructification. There is one broad distinction between lycopodiums and selaginellas, which the beginner may bear in mind with advantage. Lycopodiums have imbricated *leaves all of the same shape* spirally arranged. Selaginellas have *leaves of two sizes and slightly differing in form.* You will not readily perceive these distinctions, but look at a fresh branch of Selaginella with the aid of a lens, and you will see that between the evident leaves which stand right and left there are smaller tooth-like leaves arranged in the manner of bracts; such leaves as these you will never find in a true lycopodium. The distinctions that depend upon fructification are more subtle, and to describe them would render these pages wearisome. Better is it to quit this part of the subject at once and consider the several families named above as subjects for cultivation.

Lycopodiums.—The British Lycopodiums are scarcely worth cultivating, for the simple reason that the best endeavours have invariably failed. They are not without beauty; indeed, when we meet with a large patch of *L. clavatum*, forming a green mat two or three yards over on a tract of heath, we are compelled to admire, and can scarcely fail to be tempted by the wish to grow the plant to a similar state of perfection in our own gardens. So, again, *L. annotinum*, the most distinct of all; and *L. alpinum*, a charming evergreen tuft that may be mistaken for a savin, are well worthy of further attempts at their domestication. I must confess that I have tried them all and failed with all except *L. inundatum*, which I have had no trouble with, for it grows freely with me in the simple way I manage it, which is to bring home with me some of the soil from the bog I find the plant growing in, and having potted it in this, I keep the pot always standing in a pan of water. Plenty of fresh air it must have, or it will not last.

L. selago, the Fir-club moss, is a noble species, quite common on stony mountain tracts, a plant of great interest too on account of its medicinal properties. In the open fernery, probably, all our British species would thrive if taken up in large masses and planted in fully exposed positions in soil specially prepared to resemble that of their original sites. With them should be associated an American species, *L. lucidulum*, which closely resembles our L. selago, but is of a shining dark-green colour. This thrives in peat soil in the open fernery if favoured with shade and moisture.

Selaginellas.—Amongst these occur so many lovely forms of vegetation, that we can safely say for the guidance of the cultivator, the larger the collection the better. The delicate cushion-like growth of *S. apoda*, scarcely to be equalled by any of the true mosses, is unique for beauty. It will suggest to the reader of Milton the description of the home life of the matchless pair in the happy garden—

> "Of grassy turf their table was,
> And mossy seats had round."

The metallic blue of *S. lævigata* (*S. cæsium*) is positively marvellous, yet the plant is common and will grow in any moist warm close spot, scarcely needing warmth or daylight, though growing the better for a little of both. In *S. rubricaulis* we have the colours of the red coral and the emerald combined; in *S. formosa* we have a charming semblance to a fern, yet a kind of beauty which no fern possesses.

In practice we find them all adaptable to cool houses

and unheated cases, but to grow them in perfection warmth is essential, and they may all be treated as stove plants, and wherever there is a suitable place for them, so surely ought this class of plants to be strongly represented, for they are quite at home, and thrive where it would be next to impossible to keep any other plant in a healthy growing state, even if it could be persuaded to drag out an existence, which would not be at all likely to compliment the cultivator for the skill and trouble expended upon it. It greatly enhances the attractions of the fern house to distribute the selaginellas amongst large specimen plants, where they can have the advantage of the shade from them, putting them, of course, so as they can be readily seen, for it is no use to put a light under a bushel. They are also well suited to stand amongst strong-growing ferns, for the spreading nature of the fronds of the ferns prevents the pots being set close to each other, thus giving ample room to stand dwarf-growing plants, which require similar treatment and a deeper intensity of shade than the ferns. They will, however, grow well in a house with a suitable temperature if there are no other plants of any description but them in it, provided that the house is properly shaded. My reason for suggesting the suitability of these plants for growing between others of larger size is this—the stove is generally of limited extent, so that every inch of space is required to be made available for growing something or other, and there is always space between large specimen plants, although their fronds may meet overhead, suitable for growing dwarf plants like these, thus leaving the other space, which has the advantage of the full light, available for growing other subjects.

The best mode of growing fine specimens is in pans, for they are shallow rooting plants, and do not require a great depth of soil. By adopting pans we are enabled to give them a much larger space to spread over than would be practicable in pots. Pans of eighteen inches in diameter are a very convenient size for strong-growing kinds, whilst for the moss-like sorts of dwarf growth a smaller size is far better. The pans should have about an inch of drainage crocks broken rather fine, a layer of rough peat, and then must be filled up with a compost consisting of peat, loam, leaf-mould, and silver sand, equal parts. This should be pressed firm, a layer of sand put over it, and the cuttings laid on and pegged down. The cuttings should be good-sized pieces. I take them off at the base, close to the soil, which is better than the tops, and if they are properly attended to, they will soon take root and cover the pans. It is best to keep them renewed in this way than keep a lot of old plants; for when they are old they are bad, and get broken about, and bear no comparison to young healthy plants. The metallic lined *lævigata* can be kept in good trim by cutting it down to the pan when it becomes a bad colour, and if it has a little fine soil and sand, or sand alone sprinkled over it, and set in a warm corner, it will soon recover. Unless kept in deep shade this charming plant soon loses its exquisite colour.

Kinds which have *inæquifolia* and *viticulosa* for their type require a slightly different method of treatment in their propagation. Instead of laying the pieces on the top of the soil, the old plant should be taken out of the pan, divided into small pieces, and dibbed a few inches apart in fresh soil, in pans about nine inches in diameter, and as they cover the pans be shifted into larger sizes according to their requirements; they are slower growing than the others, and do not make large plants so quickly. Any time of the year will do for the propagation, but autumn is the best; for during the winter the cuttings get rooted if kept warm enough, and with the return of spring grow freely and soon make handsome tufts. All the kinds which are of moss-like growth, and form rootlets on their stems, are adapted to cover rocky surfaces in the fern house. Just sprinkle a little sand or peat on the rock or brick, and upon this prepared surface press a few pieces of denticulata, apoda, densa, obtusa, and others of like habit, give a sprinkle daily with the syringe, and they will soon take hold and spread and form charming little carpets of the most delicate vegetation. Specimen plants in pans will need frequent syringing when growing vigorously, but as the damp days of autumn approach, syringing must be practised less, and during winter must be discontinued altogether.

As there are few cultivators who can find room for all the Selaginellas, a selection of the most distinct and beautiful will be useful.

GREENHOUSE SELAGINELLAS.—*S. stolonifera,* green and tree-like. *S. formosa,* green, tree-like, massive. *S. microphylla,* green, slender, tree-like, red-stemmed. *S. uncinata,* blue, prostrate, wiry. *S. apoda,* green, moss-like, one of the best. *S. denticulata,* green, well known, one of the most useful. There is a white-tipped variety which makes beautiful tufts in green-house or stove. *S. Willdenovi,* green, fern-like, very hardy; one of the best. *S. lepidophylla,* dark green, like a miniature cedar tree. This is the American "Resurrection plant." *S. obtusa,* green, moss-like, beautiful.

Stove Selaginellas.—*S. rubricaulis,* red-stemmed, tree-like. *S. lævigata,* blue, a splendid climbing plant, well adapted for the fern case or to train as a climber. *S. jamaicensis,* phosphorescent, a delicate little gem.

Pepper-worts.—These plants are known in botany as the *Marsileaceæ;* they are for the most part insignificant and would have no place in this chapter were it not for the peculiar merit of one of the family which many fern-growers would like to possess. This is the *Marsilea macropus,* the Nardoo plant of Australian explorers, the plant mentioned as the last resource against starvation in the tragic story of the Burke and Wills exploring expedition. This species may easily be taken at first sight for a large-leaved oxalis, or trefoil, owing to the peculiar divisions of its leaves. It may be grown with the greatest ease in a pot of spongy peat kept constantly plunged in one or two inches depth of water. *M. quadrifolia,* a native of Germany, is also a pretty species, but it has no story to recommend it like the other.

Horsetails or Equisetums.—There is a rather troublesome weed, of very elegant structure and curious history, met with in undrained clay and loamy soils;

it is of a pale green colour, and consists of a tough and rather decumbent stem, surrounded with whorls of thread-like branches, its true leaves, if it has any, being in the form of minute scales, placed around points or rings which occur at regular intervals on the stems. The plant is known to country people as the "horse-tail" or "mare's-tail," and in botany is called *Equisetum arvense*, the field Equisetum. Though a troublesome weed, and one that is detested where it grows plentifully, it is well worth a place in the fernery, and when planted in a shady bank of peat, it spreads fast, and makes its appearance in all sorts of places, but does not drive better things out of the way, or even render itself objectionable. I have some of it in a shady part of my fernery, and very much enjoy the mixture of its elegant light green spray with such ferns as Onoclea sensibilis, and others that have bold-looking fronds. Those who know this plant, as probably most of our readers do, will be, perhaps, prejudiced in favour of the genus to which it belongs. But whether such be the case or not, I wish to recommend these plants to the notice of fern-growers, as suited to contribute in a special manner to the interest of a collection of acrogenous plants. I have all the species that are known, and one of them I consider the most elegant of all plants ever seen upon the face of the earth. This gem is called *Equisetum sylvaticum,* one stem of which is represented in the accompanying figure. If the reader can imagine a nine-inch pot, with about fifty of these stems crowded together in it, all of them arching over with exquisite grace, like feathers from the tails of birds

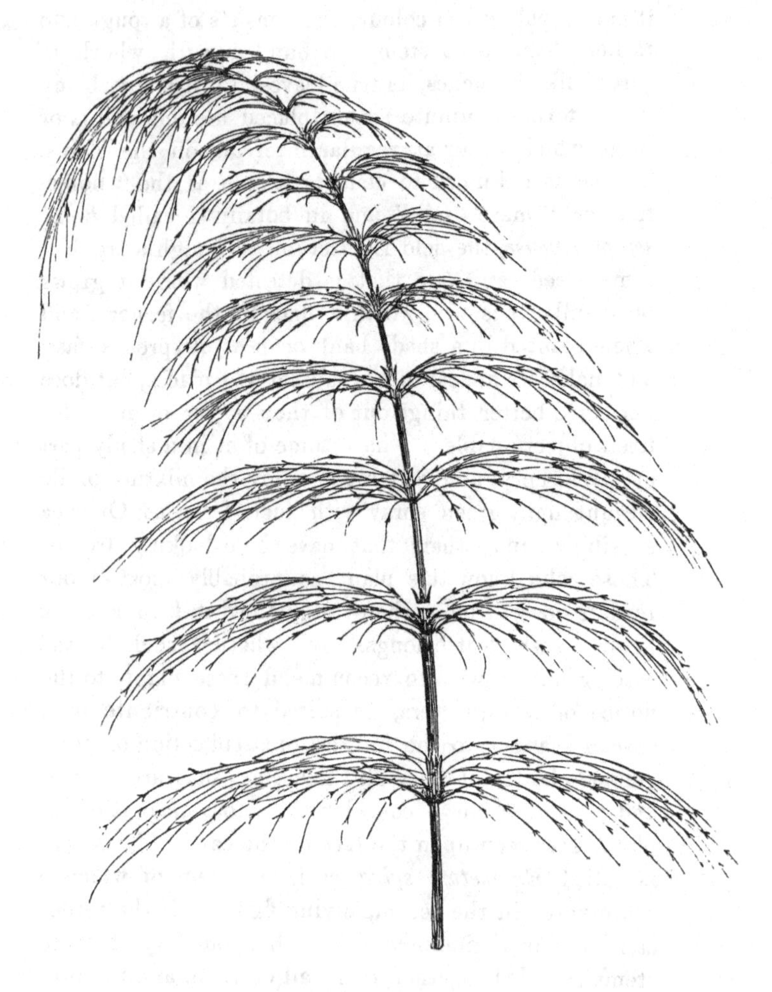

EQUISETUM SYLVATICUM.

of Paradise, the colour the most tender shade of emerald green, no apology will be needed for calling attention to it in these pages, for it is, in fact, one of the most desirable of plants for the fern garden.

Equisetum sylvaticum is a British plant, very scarce generally, but plentiful enough in some districts. When met with it is usually in a peaty soil, beside a water-course in a shady wood, or on a bank beside a ditch overhung with trees and rank herbage; always in a moist, shady spot, and if not in peat, in some light soil of similar nature. My best plants in pots are kept under a stage, and have all the drip that results from the watering of plants above them, besides the water given them in the usual way, and their appearance is so delightful, they so fascinate me that I never enter the house where they are kept without having a peep at them. They are to me a feast which never satiates, though I sometimes become tired of flowers, especially after I have for weeks constantly been visiting great gardens, and comparing and criticising bedding effects. We have it also planted out in the shadiest and dampest part of a rockery, in a cool fernery, and also in a shady part of the fernery out-of-doors. It increases fast, and may, if desirable, be parted annually in spring when it begins to grow; but to make a fine specimen it should not be parted, but be shifted to a larger and larger pot every year, and this should be done without breaking the ball when the plant is shifted; no, not even the crocks should be removed.

Another grand species is *Equisetum telmateia,* which is of more robust habit than the last, with regular whorls of branches, which differ from those of sylvaticum, that they do not branch again. This grows on dry sandy banks, and is tolerably common, especially in the southern parts of England. It grows finely in the rockery if planted in a shady spot, and though found wild in very dry positions, I have never found it succeed except in a damp position, unless assisted with frequent watering. Sandy peat is the best soil for it.

Another and most beautiful species is *E. umbrosum.* This is very distinct and very rare. The whorls of branches are rather crowded, and they all rise at a regular angle, and gracefully arch over at their ends. This grows in very shady places, and requires the same kind of cultivation.

Equisetum palustre is another exquisitely beautiful plant. By many this will be considered more beautiful than sylvaticum, for the slender branches divide and subdivide into the most hair-like ramifications; indeed, it looks as if constructed of hair, but in a manner that would be impossible to human fingers even if only in imitation of its beauty. This grows in bogs, and therefore when under cultivation must have a damp position and plenty of water.

I have also plants of *E. fluviatile,* which grows in water; *E. hyemale,* also a water plant; *E. Mackaii,* which loves moisture, and *E. variegatum,* which will grow well under almost any circumstances. But these four have no beauty. They are like rushes, tall, rigid, without branches, very pretty in a certain sense in their construction, but are likely to interest only such as are devoted to the study of these plants.

Mosses.—Though we rarely meet with these as special objects of cultivation, a large number of the most beautiful may be grown with but little difficulty in an outdoor fernery, and a few are well adapted for cool house and frame culture. Where ferns are well managed mosses are sure to appear amongst them spontaneously, and add very much to the beauty of the rockery by the tone of age and ripeness they give it. In the 'Floral World' of February, 1869, the writer of this gave his experience at length upon this subject. Those who are particularly interested in it may peruse the article with advantage.

In collecting mosses for cultivation, they should be taken with a thin slice of whatever they may be growing upon adhering to the roots. Thus obtained, they will grow freely, and spread in all directions. But when they are rudely torn from pieces of rock, the roots are injured, and the specimens suffer in consequence, frequently resulting in their death. All the mosses require a liberal supply of moisture at all seasons, to keep them in full health. A better proof of the truth of this assertion is not wanted than to point to the vigorous way in which they all grow naturally during the humid season of November and December. To keep them in first-rate condition, they should have a liberal sprinkling overhead three times a day through the summer, and at other times twice will be sufficient, unless the weather is particularly dry and warm. There need be little fear of their receiving too much. In the winter there will be sufficient atmospheric moisture, in addition to the rains, to keep them damp enough, unless it should happen to be unexceptionally dry. In that case a sprinkle overhead will be of immense benefit to them. The planting should be conducted so that each species gets a position that bears a close affinity to the circumstances in which it is found in a natural state. Those that grow on stone or brickwork should be secured to those substances, whilst those growing in damp pools and ordinary soil should have like positions allotted to them. By taking notice of the conditions under which they are found, a good idea of the treatment they require may be formed by those who know little or nothing about the subject. The cultivator will find every bit of information picked up in this way of very great service to him, at some time or other.

The undermentioned species are all particularly beautiful, and have the great merit of thriving uncommonly well upon an artificially constructed rockery:—*Bartramia fontana, B. pomiformis, Bryum alpinum, B. capillare, B. argenteum, Dicranum squarrosum, Grimmia Doniana, G. leucophœa, G. pulvinata, Hookeria lucens, Hypnum denticulatum, H. cupressiforme, H. purum, H. Schreberi, H. splendens, Leskea sericea, Racomitrium canescens, Weissia contraversa,* the principal species of Tortula, and all the Polytrichums, and Mniums.

INDEX.